GENDER. WORK, AND SPACE

Gender, Work, and Space explores how boundaries get constructed between women and men, and among women living in different neighborhoods. The focus is on work, the segregation of men and women into different occupations, and variations in women's work experiences in different parts of the city. The book argues that these differences are grounded and constituted in and through space, place, and situated social networks.

This qualitative and quantitative study of a contemporary city establishes that many women, especially those with heavy household responsibilities, are dependent on extremely local employment opportunities. Women's dependence on locally available jobs focuses attention on the existence of different employment districts throughout the city. The argument is that social, economic, and geographic boundaries are overlaid and intertwined. As employers locate firms to seek out labor with particular social characteristics, social and occupational differences are mapped in place. Neighborhood-based differences in community resources, occupational opportunities, labor processes, scheduling of work, and cultures of parenting affect the ways that families order their lives and that gender relations are enacted in daily life.

Gender, Work, and Space contributes to debates about the geography of labor market segmentation, to our understanding of sex-based occupational segregation, and, in the close attention given to the construction of social, geographic, economic, and symbolic boundaries in ordinary lives, provides a counterbalance to the focus on mobility within contemporary feminist theory.

Susan Hanson is Professor of Geography and Director of the School of Geography at Clark University in Worcester, Massachusetts. **Geraldine Pratt** is Associate Professor of Geography at the University of British Columbia, Vancouver.

INTERNATIONAL STUDIES OF WOMEN AND PLACE

Edited by Janet Momsen, *University of California at Davis*
and Janice Monk, *University of Arizona*

The Routledge series of *International Studies of Women and Place* describes the diversity and complexity of women's experience around the world, working across different geographies to explore the processes which underlie the construction of gender and the life-worlds of women.

Other titles in this series:

DIFFERENT PLACES, DIFFERENT VOICES
Gender and development in Africa, Asia and Latin America
Edited by Janet H. Momsen and Vivian Kinnaird

'VIVA'
Women and popular protest in Latin America
Edited by Sarah A. Radcliffe and Sallie Westwood

FULL CIRCLES
Geographies of women over the life course
Edited by Cindi Katz and Janice Monk

SERVICING THE MIDDLE CLASSES
Class, gender and waged domestic labour in
contemporary Britain
Nicky Gregson and Michelle Lowe

WOMEN AND THE ISRAELI OCCUPATION
The politics of change
Edited by Tamar Mayer

WOMEN'S VOICES FROM THE RAINFOREST
Janet Gabriel Townsend

GENDER, WORK, AND SPACE

Susan Hanson and Geraldine Pratt

London and New York

First published 1995
by Routledge
11 New Fetter Lane, London EC4P 4EE

Simultaneously published in the USA and Canada
by Routledge
29 West 35th Street, New York, NY 10001

© 1995 Susan Hanson and Geraldine Pratt

Typeset in Baskerville by J&L Composition Ltd, Filey, North Yorkshire
Printed and bound in Great Britain by
Biddles Ltd, Guildford and King's Lynn

British Library Cataloguing in Publication Data
A catalogue record for this book is available from the British Library

Library of Congress Cataloguing in Publication Data
Hanson, Susan.
Gender, work, and space/Susan Hanson and Geraldine J. Pratt.
p.   cm. – (Routledge international studies of women
and place series)
Includes bibliographical references and index.
1. Women–Employment–Massachusetts–Worcester.
I. Pratt, Geraldine J. II. Title.   III. Series
HD6096.W67H36   1995
331.4'09744'3–dc20   94–35111

ISBN 0–415–09940–4
0–415–09941–2 (pbk)

to Perry   *to Tohmm*

the dining room table is
yours again

CONTENTS

LIST OF PLATES

LIST OF FIGURES

LIST OF TABLES

ACKNOWLEDGMENTS

Gender, Work, and Space has evolved out of eight years of collaborative research; its genesis dates from a period when we both lived and worked in Worcester, Massachusetts. The rationale for collaboration was based, not only in our shared interest in feminism and in urban social geography but in our differences: one of us having roots in transportation and quantitative geography; the other in housing and cultural geography. These differences in focus and expertise have been pragmatically useful and have provoked each of us to learn in different ways. We wish to acknowledge and celebrate the camaraderie, pleasure, and productive work that has emerged out of a collaborative project.

We have received help from many different people and, although our names appear on the cover, we want readers to know that we could not have produced it alone. In the case of social surveys, individuals seem to get lost in the crowd, but we would like to thank all the hundreds of individuals in the Worcester-area households who, in 1987 and 1988 let us and our research assistants into their homes and shared so generously their time and their insights into their work and home lives. We are also tremendously grateful to the Worcester-area employers in the manufacturing and producer services sector, who in 1989 gave so willingly of their time and knowledge about labor market processes in their firms and in Worcester. Their candid and considered responses put a "face" on capitalist economic processes that are easily reified. We also thank the employees in these firms who agreed to participate in our study during their work day in the somewhat ambiguous circumstances of reporting on one's personal history and current employment conditions in the context of an interview carried out within the firm.

Our debts within the academy are many. We thank the many graduate and undergraduate students who over the years have given us such excellent, careful, and knowledgeable research assistance, as well as companionship, both in the field (1987–1989) and in our "home offices." Specifically, thanks to Cecile Badenhorst, Brooks Bitterman, Michael Brown, Nancy Castro, Scott Carlin, Lee Dillard, David Edmunds, Glen Elder, Caitlin Elkington, Signe Furlong, Melissa

Gilbert, Erik Hanson, Kristin Hanson, Ibipo Johnston-Anumonwo, Deborah Kahn, Ruth Katz, Tara Kominiak, Debbie Leslie, Scott MacLeod, Charlie Mather, Doreen Mattingly, Kim Miller, Sheila O'Shea, Cyndia Pilkington, Judy Pincus, Julie Podmore, Suzy Reimer, Mary Riley, Christine Salek, Jennifer Santer, Lydia Savage, Jackie Southern, Mimi Stephens, Stacy Warren, Martha Weist, and Michael Zimmer. Many of our students and colleagues have given generously of their time to read and comment on all or portions of this manuscript. These include: Trevor Barnes, Robyn Dowling, Glen Elder, Melissa Gilbert, Tara Kominiak, Doreen Mattingly, and Lydia Savage. Janet Momsen and Janice Monk, the editors of the Routledge series (International Studies of Women and Place), also contributed their timely and sage commentary on the manuscript as well as their intellectual support for this writing project. Diane LePage and Elaine Cho helped with their expert high-technology processing of text and tables. We are very thankful to them and to Anne Gibson for her cartographic assistance. Jean Heffernan, Assistant to the Director of Clark University's School of Geography, also helped our project in many ways.

Technical assistance came from other sources as well. Phil Fulton and Phil Salopek, directors of the U.S. Bureau of the Census Journey-to-Work Division, worked with us to design the special runs. John Ferrarone generously gave of his time to create photographic representations of Worcester which add significantly to our ability to convey a sense of place.

Money is not only a useful but an essential ingredient in a project such as this one, and we are most grateful for the generous support that we have received from many sources. The National Geographic Society took the initiative of funding an urban, North American project when it funded the 1987 field work. Several grants from the National Science Foundation have supported this project as well: SES 86–84347, SES 87–22383, and SES 90–22868. We gratefully acknowledge this financial assistance, which has enabled the involvement of more than three dozen students on the project. Each of us has enjoyed fellowship support at some time over the course of this research. Susan Hanson is grateful to the John Simon Guggenheim Foundation for a 1989–90 fellowship. Geraldine Pratt is thankful for the good company and writing time provided by the Centre for Research in Womens' Studies and Gender Relations at the University of British Columbia in Fall 1991. Thanks also to the Killam Foundation for a research fellowship during a 1993–4 research leave from the University of British Columbia.

This book represents our attempt to integrate a series of research projects, and some of this material has appeared in earlier research articles. We gratefully acknowledge permission to publish short passages, tables, or figures from the following: Pratt and Hanson (1988) "Gender, class, and space," *Environment and Planning D: Society and Space* 6:

15–35; Hanson and Pratt (1988) "Spatial dimensions of the gender division of labor in a local labor market," *Urban Geography* 9: 180–202; Hanson and Pratt (1988) "Reconceptualizing the links between home and work in urban geography," *Economic Geography* 64: 299–321; Pratt and Hanson (1990) "On the links between home and work: family strategies in a buoyant labor market," *International Journal of Urban and Regional Research* 14: 55–74; Hanson and Pratt (1990) "Geographic perspectives on the occupational segregation of women," *National Geographic Research* 6: 376–99; Hanson and Pratt (1991) "Job search and the occupational segregation of women," *Annals, Association of American Geographers* 81: 229–53; Hanson and Pratt (1992) "Dynamic dependencies: a geographic investigation of local labor markets," *Economic Geography* 68: 373–405; Pratt and Hanson (1993) "Women and work across the life course: moving beyond essentialism," in C. Katz and J. Monk (eds) *Full Circles: Geographies of Women Over the Life Course*, New York: Routledge; and Pratt and Hanson (1994) "Geography and the construction of difference," *Gender, Place, and Culture* 1: 5–29.

Finally, we thank our families, who provided friendship and hospitality, which sustained this transcontinental research endeavor, or, alternatively, accepted lengthy absences without question.

1

SPATIAL STORIES AND GENDERED PRACTICES

Every story is a travel story – a spatial practice.
Michel de Certeau, *The Practice of Everyday Life*, p. 115

We intend to tell spatial stories about the construction of gender differences. We write about lives in contemporary Worcester, Massachusetts and structure the plot around women's and men's labor market experiences. Our interest in paid employment does, however, take us well beyond the workplace, into neighborhoods and homes. Our argument is that social and economic geographies are the media through which the segregation of large numbers of women into poorly paid jobs is produced and reproduced.

This book is also a travel story, written about gender, place, and space, through time. We began this research in 1986, and although we attempt to draw together the segments that have been published over the last six years and reinterpret them as a whole, we would misrepresent the process involved, as well as the real movement that has occurred within feminist geography, if we were to suppress theoretical shifts and conceptual reorientations. Over time, we have chosen to address different theoretical literatures, which take varying questions as being important, and have different objectives and varying styles of representation. We see this theoretical restlessness as having positive consequences. Each theoretical tradition offers a different perspective on women's and men's lives and can be used to highlight the underlying assumptions and silences of the others. Each addresses a different audience; by speaking in multiple conceptual languages, we hope to reach different communities of readers, not just sympathetic feminist geographers, but also political economists, more orthodox economists and sociologists, and wider geographical and feminist audiences.

Perhaps most important, each theoretical tradition draws one into different spatial stories. We borrow the term "spatial stories" from de Certeau (1984), who distinguishes among different ways of representing geography, and places a higher value on some forms than on others. He

1

uses the experience of gazing from the summit of the World Trade Center in New York City as a metaphor for a detached, scientific, scopic representation of place and compares it unfavorably to the experience of walking through the city. He privileges the description of the "tour" as a less abstracted representation of geographical experience than that offered by the map. De Certeau's representational distinctions have been used to good effect in critiques of detached geographical accounts (Deutsche 1991; Gregory 1990). We are reluctant, however, to subscribe to de Certeau's hierarchies of difference; instead we appreciate the utility of both gazing from the skyscraper and walking through the street, with the understanding (which absorbs de Certeau's critique of scopic representation) that both perspectives are partial.

Different theoretical traditions encourage us to think from different vantage points. The literature on occupational segregation is, for example, largely quantitative and analysts search for generalizations across all women. Feminist cultural criticism, in contrast, often starts from the autobiographical, situated account. Geographical literature on labor markets typically gives priority to economic and structural determinants over the cultural. Different literatures start from different presuppositions and assume different points of closure. We follow Flax (1990) and Ferguson (1993) in seeing the advantage of working with multiple theoretical traditions: in their differences and complementarities they force a richer set of spatial stories. We make no claims to exhaustion; our encounters with multiple theoretical and empirical traditions remind us of the partiality of our narratives.

In this chapter, we provide an overview of these theoretical traditions, to show how each contributes to an understanding of the spatial construction of gender relations. We deal with them more or less in the order with which they have influenced our thinking, beginning with the economic, sociological, and feminist geographical literatures on occupational segregation, moving to feminist and industrial geographers' treatments of occupational segregation and labor market segmentation, and finally to theorizing about gender identity and space within contemporary feminist theory.

OCCUPATIONAL SEGREGATION – WHAT'S SPACE GOT TO DO WITH IT?

It is unexceptional for women in most parts of the world to hold jobs outside of the home, even when we have the responsibility of mothering very young children. In the U.S. in 1990, 57 percent of women between the ages of 16 and 64 were in the waged labor force, as were 59 percent of mothers of children under age 6. Women constitute a significant proportion (45 percent in 1990) of all waged workers (U.S. Bureau of the Census 1990). Similarly, in 1990, 58.6 percent of married women in the U.K. and 57.9 percent of married women in Canada were in the paid work force (McDowell 1991a; Statistics Canada 1993).

Women do, however, continue to participate in the labor force in ways that differ from men. One point of difference is that women as a group tend to earn less. At every educational level, women in the U.S. earned less in 1990 than men with the same amount of schooling, and although the pay gap between women and men who work full time has declined in recent years, it continues to widen with age. In 1991 college-educated women between the ages of 25 and 34 earned on average 90 cents for every dollar earned by a man of similar age and education, an increase of five cents on the dollar since 1981 (Steinbruner and Medoff 1994). The 1991 wages of college-educated women between 45 and 54 years of age, however, were only 62 cents for every dollar earned by men of the same age and education, the same amount earned by male high school graduates (Medoff 1994; for comparable 1986 data, see Reskin and Roos 1990: 20).

Some may interpret these data as good news, evidence that gender inequality is declining over time and is especially on the wane among younger age cohorts. The compression of the gender wage gap is due in part, however, to a decline in men's wages as well as to the increase in women's (Steinbruner and Medoff 1994). And the wage differentials between the age groups reflects the fact that most women enjoy rather less in the way of wage increases over their lifetimes, than do most men (Rosenfeld 1979). The enduring gender wage gap reflects not only different rates of career mobility within particular occupations but also the fact that women and men tend to work in different occupations and industrial sectors. For example, in 1980, 48.9 percent of employed American women (compared to 16.1 percent of men) worked in administrative support (including clerical) and service occupations, while 28.5 percent of all men and 3.3 percent of women worked in precision production and craft, and transportation occupations (Reskin and Roos 1990: 4). Typically the sectors and occupations filled by women tend to have lower income ceilings, poorer benefits, and less career mobility.

The tendency for women to have different occupations from men is what is referred to as occupational segregation; it is a phenomenon that appears to be remarkably persistent. Historical studies (Gross 1968; Beller 1984; Jacobs 1989) indicate that levels of occupational segregation in the United States were essentially constant from 1900 to 1960. During the 1970s sex-based occupational segregation did decline slightly; even so, in 1980 almost half of all American women and 53 percent of men worked in occupations that employed, respectively, at least 80 percent women and men (Rytina and Bianchi 1984). Reskin and

3

Roos (1990) have made a close study of the 33 male-dominated occupations into which American women seemed to have made significant inroads during the 1970s, increasing their representation by at least 9 percent. On closer examination, they find very little evidence for genuine integration. Women are either ghettoized into low-status specialties, low-paying industries, and part-time work or the occupation has been re-segregated as predominantly female (e.g., this is the case for insurance adjusters, examiners, and investigators). "The answer to our question of whether occupation-level desegregation translated into job-level integration is a resounding no. Women and men tend not to work alongside one another on the same jobs in these newly desegregating occupations' (Reskin and Roos 1990: 72).

Opinions about whether occupations should be reconstituted so that they are filled equally by women and men, and how this might be accomplished, depend on one's analysis of the sources of occupational segregation. Much of the debate about the reasons for continuing occupational segregation can be read as being implicitly geographical, insofar as the argument turns around where the causes are located: in the home and within the family or through workplace-based practices.

Occupational segregation begins at home

A range of theorists locate occupational segregation within the home, although they vary quite radically in their normative and political stances. Those who explain occupational segregation and gender-based wage differentials in terms of human capital theory (e.g., Mincer and Polachek 1978; Mincer and Ofek 1982; Polachek 1981) tend not to see either of these labor market outcomes as being particularly problematic (or at least, not as signaling a structural problem within the labor market). They locate women's labor market position in terms of their household responsibilities and interpret a woman's decision to work in a female-dominated occupation as a rational choice, made within the nexus of these additional work obligations. The rationale for this decision hinges on a woman's expectation of moving into and out of the labor force as the demands of childbearing and child care ebb and flow. According to this line of reasoning, a woman's expectation that she will spend fewer years in the labor force over the course of her life than will a man makes it irrational for her to spend many years in pre-job training. Additionally, human capital theorists argue that traditionally female occupations are less likely than are male occupations to have apprentice-type training periods built into the early years on the job, during which time wage levels are suppressed. They argue that female occupations thus have lower "start-up" costs, an attractive

4

feature to an individual who intends to move in and out of the labor force.

Human capital theory has received many challenges, both empirical and theoretical, which we agree with and review in Chapter 5, but it should be recognized that some feminist attempts to explain occupational segregation also locate its origins within the home, in women's heavy domestic workloads (without assuming the rationality of individual choices or drawing the same normative conclusions). This was particularly true of feminist accounts in the late-1970s and early-1980s. Participants in the domestic labor debate, for example, interpreted women's position almost exclusively in terms of their domestic responsibilities, seeing women's provision of this labor as being functional to capitalists insofar as it cheapens the costs of social reproduction and indirectly dampens wage demands. The Marxist and socialist feminists who were engaged in this debate disagreed about whether the burden of domestic work was thrust upon women by capitalists or by men as patriarchs, that is, they disagreed about the relationship between capitalism and patriarchy, but there was general agreement that subordination within the home in large part explained women's subordination in the labor force. (For a review of these arguments, see Tong 1989.) It is worth noting, however, that domestic labor theorists largely draw out the temporal rather than the spatial relations between home and work, as, for example, in research that emphasizes the implications of women's heavy domestic workloads for their types of waged employment (Hartmann 1987). Our interest in geography leads us to complement these temporal stories with spatial ones.

Occupational segregation begins at work: labor market segmentation theory

In the early-1980s many feminists attempting to explain occupational segregation turned their attention away from the household and the individual choices made in the context of familial responsibilities, toward the workplace, in line with a more structural reading of segmented labor markets (Game and Pringle 1983). From England and Farkas (1986: 154) comes a strong statement that removes the process of occupational segregation from the home: "women's domestic responsibilities cannot explain their absence from most male-dominated jobs." Others (e.g., Reskin and Roos 1990; Walby 1986) are equally adamant that women's family roles are largely irrelevant to their occupational choices and that the origins of occupational segregation must be sought within the labor market.

Segmented labor market theorists see the labor market as being divided into different segments, not as the result of rational choices on

5

the part of employees and inequalities in the attainment of human capital that are structured outside of the labor market, but as the result of processes internal to the labor market itself (Morrison 1990; Peck 1989). Segmentation theorists cite a diverse range of causes for segmentation, including technological requirements, product market instability, employer discrimination, and employers' attempts to control labor and labor costs through divisive practices. In an idealized conception, the labor market is divided into internal (or primary) and external (secondary) segments. In the internal segment, wages and job mobility are determined by institutional rules, often negotiated by employers and unions, and not determined exclusively through market forces. Workers in the internal or primary sector tend to earn better wages, enjoy more job stability and better working conditions, and have more promotional possibilities than do those in the external or secondary segment. There is little mobility across sectors, and workers in the secondary sector effectively become trapped within it. Central to segmented labor market theory is the point that workers' labor market preferences are shaped within segments of the labor market itself, not outside of the labor market as is posited in neoclassical accounts (Lang and Dickens 1988). (For reviews of segmented labor market and human capital theories of occupational segregation, see Morrison 1990; Peck 1989; Pinch 1987; Redclift and Sinclair 1991.)

Segmented labor market theory has been criticized because of its relative inattention to the processes that lead certain categories of workers into different segments, namely white males into the primary segment and women and visible "minorities" into the secondary ones (Redclift and Sinclair 1991). A number of theorists have, however, used segmented labor market theory as a framework, into which they insert the dynamics of patriarchy at work. Explanations for women's exclusion from the primary segment build on the sexist practices of male employers and employees. Male employers may be reluctant to hire women for the most prized jobs because of gender stereotypes, worries about complaints from male employees, and their more general fears about losing male advantage: "Dispensing with sex differentiation at any level threatens it at every level, so ignoring sex in the labor queue challenges the advantaged positions of male employers and managers" – both inside and outside the workplace (Reskin and Roos 1990: 37). White, male employees also have organized through unions and professional organizations to shelter jobs for themselves (Cockburn 1983; Reskin and Roos 1990; Rubery 1978; Walby 1986).

Although the debate about how best to explain occupational segregation has been implicitly geographical in the sense of attending to where the locus of causation lies, until recently few have explored this geography or have drawn connections between relations within the home and

6

the workplace in other than temporal terms. There has been an unproductive tendency to claim the importance of one side of the duality–either home *or* work. Peck remarks in reference to early segmentation models: "[t]he appreciation of the supply-side of the labor market in these models was not particularly sophisticated" (1989: 45-46). Later segmentation theory, associated with the Cambridge labor market segmentation school, has placed more emphasis on labor supply (Peck 1989; Morrison 1990), but, in general, the connections between different spheres of life remain relatively undeveloped, and when connections are made, the geography of the relations between home and work tends to be ignored.

Does occupational segregation begin with networks?: economic sociology

One emerging strand of thought that *has* insisted on the inseparability of economic and social realms of life (though not in the context of gender-based occupational segregation) directs attention to what Granovetter (1985) calls embeddedness – the embeddedness of individuals in networks of association and the embeddedness of economic activity in social life. Referred to as economic sociology, this work speaks not of human capital, accumulated by atomized individuals, but of social and cultural "capital," forms of noneconomic knowledge that affect the economic practices of individuals and groups as well as access to particular occupations (Granovetter 1985, 1988; Fernandez Kelly 1994). Networks of personal contacts and the information, norms, and values that are filtered through them do not respect the neat divisions scholars have drawn between the economic and the social, between production and reproduction, between work and home; they are the medium through which individuals develop aspirations and, often, acquire access to jobs.

Of distinct importance are the type and quality of resources that individuals in different networks of social relations command: Are the people in one's social network relatively similar (to oneself and to each other), or are they diverse? What range of experiences have they had and what kind of information do they have access to? Are they well-connected to people in other networks with different characteristics? Granovetter and Tilly (1988) have argued that understanding personal networks is central to understanding labor market inequalities. They see these inequalities as resulting from bargaining and conflict among employers, workers, households, organizations, and government. In their view, personal networks and the resources they represent are crucial to the outcome of this conflict (see also Manwaring 1984; Blackburn and Mann 1979).

Although the term "embeddedness" has (to us, at least) a spatial connotation of rootedness in place, much of the discussion within

7

economic sociology has not explicitly exploited the geographic dimensions of this concept. A recent, and fascinating, exception is Fernandez Kelly's (1994) study of teen pregnancy and motherhood in the Baltimore ghetto, in which she explores the ways in which teenagers' access to information and cultural resources depends on physical and social location – in this case, the neighborhood. Our Worcester study weaves together the insights of economic sociology, of labor market segmentation theory, and of geography to understand how different structures of employment opportunity are constructed in place.

Geographers have long been attentive to the spatiality of urban life, although the connections they have drawn between urban form and women's disadvantaged position in the formal labor market have, until recently, remained either implicit or highly generalized. Early work by feminist geographers tended to focus on the friction of distance: on gender differences in travel-activity patterns and on the spatial separation of residential and employment districts. More recently, economic geographers have paid closer attention to localities and to variations in labor market segmentation across places. Each tradition tends to tell different spatial stories, ones that could inform each other in productive ways.

Commuting patterns and women's waged labor

In the mid-1970s, feminist geographers began to make considerable effort to understand how spatial constraint enters, in a very general sense, into processes of occupational segregation (for reviews, see MacKenzie 1989; Bowlby *et al.* 1989). A spate of time-space geographies and journey-to-work studies undertaken in the 1970s and early 1980s documented the spatial constraints that women experience. These studies revealed that women tend to travel less frequently, over shorter distances, and via different means of travel than do men. They also explored the implications of these travel patterns for access to services and paid employment (Tivers 1985; Hanson and Hanson 1980, 1981; Forer and Kivell 1981; Madden 1981; Howe and O'Connor 1982; Pas 1984). Gradually a direct link between commuting time/distances and occupational segregation emerged from this literature with the finding that women in female-dominated occupations have shorter worktrips than other women (Hanson and Johnston 1985; Singell and Lillydahl 1986). More generally, MacKenzie and Rose (1983) argued that the spatial separation between residential suburb and urban workplaces is integral, not incidental, to the conceptual and practical separation of home and economy and to the difficulties that women experience in combining domestic and wage labor: women located in suburban

8

residential areas are unlikely to travel the long distances required to get to urban workplaces.

Because our research emerged out of and extends this tradition, it is important to reflect on the characteristics of the spatial stories that are associated with it. Geography has tended to be read, within this strand of feminist geography, as distance and separation, highlighting the separation between home and work and the distance to services and well-paying jobs. In addition, this work focused primarily on gender differences and tended to submerge the differences that exist among women. Only relatively recently has attention been given to variations in home–employment distances among different groups of women (Johnston-Anumonwo 1992, 1994; McLafferty and Preston 1991, 1992). With access to different housing and labor submarkets, the experiences of African American and Latina women, for example, are often very different from those of white women.

This early work also paid insufficient attention to differences in the practice of occupational segregation across places and how these practices are constituted through and in place. In many cases, women's and men's lives have been read with very little sensitivity to the context, or to the variations across the contexts, in which they are situated. In this respect, early feminist geographers' work on occupational segregation was vulnerable to critiques of spatial science, namely that the friction of distance is "ripped out of its social contextuality and modeled as a quasi-Newtonian 'independent variable'" (Soja 1989: 149). G. Rose (1993) argues that this reflects the fact that feminist geography has been conceived in opposition to, but within the frame of, masculinist geography. As a more empirical variation of this critique, Kim England (1993) has urged that feminist geographers refocus their attention away from the friction of distance per se (e.g., women's generally short journey to work) to a fuller appreciation of a multiplicity of contexts and roles.

In both drawing upon and moving on from this earlier work of feminist geographers, we call attention to two points. First, it is worth reiterating that in the sociological and economic literature on occupational segregation, the friction of distance – the distance and connections between home and employment – tends to be ignored altogether, as are the specificities of context.[1] Certainly it is assumed that aspatial social and economic processes are geographically invariant, at least across industrialized countries: "labor market theories are deemed to be portable from place to place" (Peck 1989: 42). The truth of this assumption seems questionable, given, for example, differences in female participation rates across regions (Moseley and Darby 1978; Stolzenberg and Waite 1984; Walby 1986; Ward and Dale 1992) and studies that demonstrate that gender-typing of occupations varies from place to place (McDowell and Massey 1984; Parr 1990). Second, we wish to move

9

beyond the construction that sets up "friction of distance" and "context" as the poles of a dualism. It seems inappropriate to pose the two as a polarity when in fact it is the friction of distance that directs attention to the importance of context, insofar as the spatial constraints experienced by many women suggest that many women are extremely dependent upon local employment opportunities and other resources.

The connections between the two poles became more obvious to feminist geographers as the "localities" tradition emerged in industrial geography. Localities research suggests that not only is it necessary to embed distance within a fuller set of socio-spatial relations in order to understand the occupational segregation of women, but also the friction of distance and gendered divisions of labor are crucial to the constitution of places and local economies. That is, the meanings and impacts of distance are not only place-contingent and socially constituted; because individuals are constrained by space, different ways of life develop in different places.

Localities and labor market segmentation

The division between economy and family has been almost perfectly encoded by subdisciplinary boundaries within geography; traditionally, economic geographers have virtually ignored gender differences, leaving such concerns to social and urban geographers (who have in turn replicated the same divisions between economy and family within their own analyses, even in descriptions of residential landscapes, for example within social area analyses; for elaboration of this critique see Pratt and Hanson 1988; G. Rose 1993). Within contemporary discussions of economic restructuring, Christopherson (1989) notes a masculinist bias, with, for example, an excessive amount of attention dedicated to restructuring within manufacturing industries, excessive because most job growth and much job restructuring has occurred within the service sector. Christopherson argues that the marginalization of sexual difference as a theoretical and substantive focus is especially problematic because the gender division of labor is part of the industrial restructuring process, rather than incidental to it. One cannot, for example, understand the growth of contingent employment (subcontracting, temporary, and part-time employment, much of which is taken up by women) without situating those labor practices within the existing gender division of labor.

This point certainly has not been lost on all industrial geographers; in fact it was central to Doreen Massey's relatively early work on regional restructuring in the U.K. Massey (1984) argued that processes of economic restructuring and deindustrialization have to be read at regional and intraregional scales and not simply at a national one.

10

Rather than a spatially undifferentiated process of deindustrialization, a much more complex and variegated process of deindustrialization and reindustrialization was occurring in the 1970s and 1980s, with certain types of industries moving away from cities in central regions into suburban areas and peripheral regions. The attractions of different localities are variable, depending on previous economic activity, local cultures, and traditions of labor organization. The geography of gender relations was, she argued, integral to industrial relocation. For example, many low-wage branch plants employing women moved into old heavy industry and mining areas where traditional jobs for men were declining. A strict sexual division of labor between breadwinner and homemaker had been the norm in these areas. Women constituted a reserve of "green" labor, most having had no experience with waged labor and labor organization; many branch plant employers found women to be an attractive labor supply (see also Cooke 1983).

Further, the effects of increased female waged labor on gender relations were felt differently in these old industrial areas than in ones, such as Cornwall, where female participation rates were also rising but where male unemployment was not as high. In the former, many men felt immensely threatened by the loss of traditional masculine jobs and were torn between rejecting new forms of employment as "girls'" jobs and resenting the fact that young women were taking these jobs while older men were left unemployed. In Cornwall, conflict and debate also moved around familial ideology, but in this place the concern was that women were forced to drive fairly long distances to take factory jobs in new industrial parks and not the fact that women were engaged in paid employment. As Massey tells it, in Cornwall the major conflict took place, not between working men and women, but between different factions of capital – local and national – which were competing for low-cost female labor. Thus in different areas the fact of increased female labor force participation generates different stresses, which draw in different actors.

Massey's early work was important because it was both geographical and gender-sensitive. She drew attention both to connections between economic restructuring and gender relations and to variations in gender and work relations across localities. Still, critics of this work suggest that it was flawed precisely because she failed to problematize sex-based occupational segregation (Walby and Bagguley 1989; Bowlby et al. 1989). Walby and Bagguley argue that Massey (and Cooke) treated the sex-typing of occupations as a constant, insofar as they assumed that women's participation in the labor force was contingent on the expansion of certain low-skilled "female" occupations and industrial sectors. Critics also argued that clerical occupations and the service sector are almost entirely ignored in these accounts and that women's traditional

11

roles tend to be seen in terms of the conditions of men's jobs. For example, Massey explained the low incidence of paid employment among women living in the coal towns of South Wales almost entirely in terms of the circumstances of men's employment: the fact that men's work was done in shifts and therefore demanded large inputs of domestic labor. But Walby and Bagguley (1989) note that women worked in the mines until they were prevented from doing so by the 1842 Mines Act; clearly only a more complex and detailed account will explain why women were, until recently, excluded from paid employment in the mining towns of South Wales.

Critics also argue that, although Massey (1984) and Cooke (1983) mentioned gender as a factor, they continued to imply that gender relations are a by-product of capital or capital–labor relations. Certainly, one is struck by Massey's discussion of the debate in Cornwall about the best location of women's waged employment: in distant industrial parks (favored by the representatives of national capital), as opposed to close to or even in the home (favored by locally based capital). Massey notes that this tussle over economic development strategies is an argument about spatial form "in which the ideology of the family is invoked, but which is actually about who taps a [low-cost, female] labor supply and how" (1984: 231). What Massey presents as an intra-class conflict (between national and local capitalists) masqueraded through familial ideology was probably a much more complicated struggle, in which both patriarchal and class relations were being redefined. Local capitalists, in wishing to maintain close ties between home and work for their wives and daughters, may have been acting as both patriarchs and capitalists; we lose sight of this dynamic when the story is abbreviated to one about class relations.

Massey's approach can be complemented and enriched through a dialogue with the literature on occupational segregation (and vice versa). Walby and Bagguley (1989) recommend dual systems theory as a way of bringing the two literatures into contact, referring to an approach that sees patriarchy and capitalism as two separate but inter-acting systems (Hartmann 1979). While Walby (1989) has attempted to bring more subtlety to the concept of patriarchy, a term that many feminist theorists now feel more comfortable using as an adjective than as a noun (Barrett 1980), the use of the systems metaphor does imply a level of abstraction and a degree of closure that is at odds with recent epistemological tendencies toward nontotalizing accounts.

Our preference is to explore the linkages between male domination and class relations more locally and contextually (rather than at an abstract systems level); indeed, one of our contributions to and beyond the localities approach is to work at a much finer geographical scale than is typical within this tradition. A locality is usually defined as a local labor market, the area encompassed by the daily commuting distances of employees, and this is almost inevitably conceived and operationalized at a metropolitan scale (see, for example, Bagguley et al. 1990).[2] The results of numerous studies documenting that women typically commute very short distances to paid employment suggest that this traditional definition of spatial labor markets, and by extension localities, is rooted in men's experiences, particularly those of white, middle-class men. The commuting ranges of many (not all) women tend to be smaller than those of many men, creating a number of separate labor markets within any single large metropolitan area. This suggests that the geography of gender relations is constructed at a very local scale and that the experiences of being in a gender and of occupational segregation may be different from place to place, even within a metropolitan area.

What the links might be between commuting distances and labor market segmentation is a matter of some debate. Peck (1989) downplays the relationship when he draws on segmentation theory – aspatially conceived – to criticize the tendency to define local labor markets in terms of commuting distances:

> Segmentation can be seen as a set of processes that "slice up" local labor markets, undermining their internal coherence to a potentially debilitating degree. Workers in *different* segments of the *same* labor market may share little in the way of common experiences of employment (or unemployment). Indeed, workers who occupy completely different segments of the labor market (and hence do not compete with one another for jobs) are to be found in the same workplace (e.g., managers and canteen workers). They may also be found living in the same house (e.g., husband and wife). . . . The notion of the friction of distance is, then, not at all adequate in isolation as the basis for explaining the labor market behavior of such groups.
>
> (Peck 1989: 49)

To understand the links between labor market segmentation and space, Peck draws attention away from commuting distances and questions of spatial access within local labor markets, to the ways that state policies, and labor supply and demand processes come together in different ways in different places, so as to create locally variable patterns of labor market segmentation.

While Peck is no doubt correct in arguing that the friction of distance is alone insufficient to understand access to particular jobs and labor market segments, he possibly overstates the "debilitating" incoherence of local labor markets by ignoring the spatial fragmentation of employment within metropolitan labor markets, which may entail the clustering of jobs from different occupational segments. Peck's observation that labor market segmentation cannot be reduced to a simple spatial concept of friction of distance should not lead us to ignore the ways in which the friction of distance enters into the process of labor market segmentation *within* urban areas.

The importance of intraurban labor markets has been stressed by some economic geographers, especially Scott (1988), and Clark and Whiteman (1983).[3] Scott interprets his empirical work on animated film studios in Los Angeles as "dispel[ling] forthwith any notion that metropolitan areas invariably constitute the minimum geographical level of local labor market differentiation" (Scott 1988: 150). He has continued to explore the relations between urban form and labor market segmentation in the electronics assembly industry in Southern California (Scott 1992a, 1992b).

Our work complements and extends that of Scott in a number of ways. First, we focus more explicitly on gender. In his more recent work, Scott (1992a) has begun to unravel the ways in which employers' perceptions of skills interface with the gendering and racialization of task assignments in the Southern California electronics assembly industry. It is arguable, however, that Scott's earlier treatment (1988) of gender is susceptible to some of the criticisms that have been leveled against segmentation theory, namely the tendency to collapse gender into race, ethnicity, and occupational status by treating the situations of female workers as interchangeable with those of various ethnic and racial minorities and "low-skilled" workers. In his discussion of commuting patterns, for example, he describes the familiar distance decay of worktrips, with fewer workers traveling longer distances. "These relationships however, are much influenced by the socioeconomic status of commuters." He then clarifies this point with an illustration, in which socioeconomic status is gendered: "In the case of female factory workers, for example, the commuting shed typically covers a comparatively restricted spatial area; but in the case of male managerial labor, it may range over the whole metropolitan region" (1988: 121). Numerous studies have shown that the commuting sheds of many female workers are constrained relative to those of many men, but by collapsing gender into socioeconomic status, Scott suppresses the dynamics of gender and space.

It is also the case that economic geographers have tended to highlight class relations rather than gender relations as causal forces. Scott (1988) notes, for example, that technical workers in the animated film industry live closer to work, have lower wages and lower job status than rendering artists and creative workers, and have jobs that are especially likely to be subcontracted overseas. He explains the commuting patterns and sub-contracting arrangements solely in terms of production-related factors:

14

wages and skill level, respectively. The fact that technical workers are mostly women and the other animators seem to be men may not, however, be irrelevant to the story: women tend to work closer to home for reasons that go beyond wages, and women's position within the unions that fight subcontracting arrangements typically has not been strong, for reasons that include gender politics as well as production relations.

Second, our work complements the work of Scott by focusing more comprehensively on the relations within households and communities to consider how they affect paid employment. By looking inside households, communities, *and* workplace establishments we have tried to draw together the insights of both feminist and industrial geographers to understand the "dynamic dependencies" between employers and employees in creating local labor markets and labor market segmentation. In his critical review of the literature on space and segmentation theory Morrison (1990) argues that conceptual "progress" will be stymied until a more rigorous distinction is made between regional labor markets (defined by commuting distance) and local labor markets (defined from the perspective of the enterprise). While Morrison's point may be useful, one of our central arguments is that it is through understanding the empirical fusing of these two categories that one can understand the creation and sustenance of intraurban labor markets.

Third, our vantage point differs from that of Scott insofar as we focus on employment patterns and processes, broadly conceived, throughout a metropolitan area rather than within a particular industry (or set of industries). By examining the full range of employment opportunities in one place, we have the advantage of being able to explore variations in labor market segmentation across all types of waged work throughout the metropolitan area.

In sum, the spatial stories about the gendered construction of labor markets told by a small number of industrial geographers are important because they draw attention to variations across places and to the ways that gender divisions of labor partially constitute locally distinctive labor markets. Still, they are limited by the facts that the processes that lead to sex-based occupational segregation are not thoroughly examined and that the experiences of women are sometimes collapsed into a caricature of "green labor." In homogenizing the experiences of women (and sometimes combining them with those of all visible minorities), these accounts share a weakness with the literature on occupational segregation; they both tend to treat the category "woman" unproblematically, and differences among women are only superficially explored. This weakness has been recognized recently by feminist geographers (Bondi 1992; McDowell 1991b; G. Rose 1993), in part, through a critical engagement with feminist theory.

15

FEMINIST APPROACHES TO SUBJECTIVITY AND SPACE, AND SITUATED THEORIES

Our reading of feminist theories has broadened since we began, as is the case for many feminist geographers.[4] Recent feminist theory poses some interesting challenges to a reinterpretation of both the theoretical traditions outlined above and our empirical research in Worcester. We make no attempt at a systematic or exhaustive review of contemporary feminist theory as it pertains to spatial relations, confining the discussion, instead, to a consideration of how contemporary feminist theory and the Worcester case study can be made to enact upon each other.[5] We consider first two types of spatial stories that contemporary feminist theorists tell about subjectivity and femininity: one about containment and another about mobility. We then trace the influence of contemporary feminist theorizing about essentialism on our own work, and, finally, signal the way that feminist theory has forced us to pay closer attention to the positionality and partiality of our interpretations.

Containment stories

Feminist scholarship is rich in stories that interpret feminine subjectivity in terms of proximity and enclosure. For example, tropes of proximity run through psychoanalytical accounts of feminine psycho-sexual development (Doane 1987; Game 1991; Jardine 1985; G. Rose 1993; Tong 1989). Another set of spatial stories clusters around the spatial restrictions in feminine bodily comportment (Young 1989; Bartky 1988).[6] Griselda Pollock (1988) maps, for example, the theme of restricted spatial mobility and women as objects of "the gaze" onto her reading of various representational practices and urban space in the late nineteenth-century city. She describes how the canonical paintings of male modern artists such as Monet and the writing of poets such as Baudelaire similarly assume a male audience and present women as the objects of the male "gaze." This was not simply a sexist practice (although it was that too): "paintings of women's bodies [are] the territory across which men artists claim their modernity and compete for leadership of the avant-garde" (Pollock 1988: 54). Women's bodies were the medium through which class and gender relations, and cultural movements, were represented and reworked.

Pollock's work is especially interesting to geographers because she places women's experience and representation of space within a historical and geographical context. She links cultural representations to gender and class relations in modern societies and to concrete lived experiences in what she terms the modern spectacular city, in which the ideological divisions between public and private were concretized in urban form and the public central city became a highly eroticized zone where men of all classes mixed with working-class women, but from which middle-class and bourgeois women were excluded (see also Wilson 1991). The exclusion of nonworking-class women as active viewers of visual art is but one instance of a more pervasive exclusion from public life. The inclusion of working-class women as objects within cultural representations was a replaying of their very partial manner of inclusion in the public life of the city.

Pollock is keen to move beyond a simple reflection theory of culture, however, and she makes the important point that the canonical paintings of modern art not only reflect societal constructions of femininity; they actively shape them. "Instead of considering the paintings as documents of this condition, reflecting or expressing it, I would stress that the practice of painting is itself *a site for the inscription of sexual differences*" (1988: 81).

We can draw upon these types of feminist containment stories to situate our study (and much of the work of feminist geographers) insofar as we can interpret our study as continuing the theme of containment at another scale and in another context, that of urban neighborhoods and intraurban labor markets in contemporary Worcester, Massachusetts. Feminist containment narratives also provoke us, however, to tell a richer set of spatial stories. Without masking the substantial theoretical divergences among even the few examples cited above, and within contemporary feminist theory more generally (Hirsch and Fox Keller 1990), we take two general lessons from recent accounts of femininity and spatial containment.

First, subjectivity is conceptualized as a process, continuously inscribed and reinscribed through discourses, cultural representations, and everyday practices. It is inscribed in and through, among other means, urban form. This reading of subjectivity differs from that which seems implicit in earlier feminist geographical work and opens up a richer set of relations between subjectivity and space. Implicitly, much feminist geography has seemed to rest on what Grosz (1992) identifies as a causal model for conceptualizing the relations between subjectivity (and the body) and urban space. Space and places are often viewed as constraints or resources that prevent or enable women to actualize their potentials. This interpretation would seem to rest on an implicitly humanist version of the subject, with individuals portrayed as at least partially autonomous agents with the potential to conceive of themselves and their desires apart from their environment. A processual reading of subjectivity suggests a different model. "What I am suggesting is a model of the relation between bodies and cities which sees them, not as megalithic total entities, distinct entities, but as assemblages or collections of parts, capable of crossing the thresholds between substances to

17

form linkages" (Grosz 1992: 248). So, for example, Iris Young is willing to speculate about the interface between feminine subjectivity and the micro-geography of bodily comportment: "I have an intuition that the general lack of confidence that we [women] frequently have about our cognitive or leadership abilities is traceable in part to an original doubt in our body's capacity" (1989: 67).

This model of the reciprocal constitution of subjectivity and geography prompts us to think more fully about the creation of different gendered and social identities in different places and to look at and listen carefully for the many different ways that these identities are inscribed. It pushes us beyond considering geography and local labor markets only in terms of opportunities and constraints, as envelopes of resources that allow or disallow individuals to fulfill their preconceived potentials; it opens the recognition that gendered identities, including aspirations and desires, are fully embedded in – and indeed inconceivable apart from – place and that different gender identities are shaped through different places. For example, as we show in Chapter 7, varying types of male-dominated occupations are open to women living in different areas. Accessibility to different traditionally male occupations conditions the experience of being a woman, and this occurs differently in different places. So too, we argue that employers' perceptions about and responsiveness to class preferences about desired hours of work leads to variation between middle-class and working-class areas in the scheduling of part-time and shift work. As different traditions of scheduling take root in different areas, different forms of family life, including gendered work habits and expectations, develop.

As a second general point, we want to register Griselda Pollock's observation that the representation of women's bodies by male painters was not only about masculine sexual pleasure but also about competition for leadership of the avant-garde. This would seem to instantiate Foucault's (1990) argument that sexuality has become an especially dense "transfer point" for relations of power since the seventeenth century because so many discourses that mediate and control class and other relations center on sex. This leads us to look closely at how discourses about different social relations fold into each other, how, for example, discourses of family regulate employment practices or those centered on women's capabilities as workers reflect and reproduce racial difference.

Our vantage point, as geographers who have entered the study of gender through labor markets, also stands to enrich the literature on embodied subjects. In Paul Smith's judgment (1988), much of contemporary debate about subjectivity (both within and without feminism) has been carried out in relation to three cultural practices: film, television, and literature. No doubt a reaction to various forms of economism,

18

especially Marxist versions, this is nevertheless a very selective reading of "culture" and "the subject," which tends to reproduce ideological divisions between home and work, consumption and production. Our interest in the traffic between symbolic and concrete spaces also enriches the spatial stories told about embodied subjects, by tracing out how symbolic spatial stories become concretized in urban spaces, with very real implications for the women and men living within them. We note in Chapter 6, for example, how the reputation of an area as dangerous dampens the labor supply of (white, middle-class) clerical workers and alters employers' labor practices to include the extensive use of temporary workers.

Recent feminist scholarship has been especially attentive to the multiplicity of social relations that structure individual's identities in interdependent and contradictory ways. The experience of being a woman is very different depending on how one is positioned in terms of class, race, religion, sexual orientation, colonialism, etc. This recognition has tended to be articulated through another spatial story, that of mobility. We turn to consider how our material evidence from Worcester can be made to intersect with this feminist spatial story.

Mobility stories

The metaphors of exile, nomadism, and the idea of a continuous shuttle between center and margins, have been used by feminist theorists (as by many other cultural theorists) to articulate a type of consciousness and cultural condition. They convey a sense of a subjectivity that is in process, unstable and interpellated by diverse and sometimes contradictory subject positions. Trinh Minh-ha describes this consciousness through the image of "a permanent sojourner walking bare-footed on multiply de/re-territorialized land" (1990: 334). Drawing, in part, on Zen teachings, she writes, "*If you see the Buddha, kill the Buddha!* Rooted and rootless . . . walking on masterless and ownerless land is living always anew the exile's condition" (1990: 335). For Trinh, movement and exile are metaphors that articulate an attempt continuously to displace boundaries between center and margins, thereby displacing controlling reference points. Gillian Rose (1993) has labeled this space that "simultaneously grounds and denies identity" as paradoxical space. "This geography describes . . . subjectivity as that of both prisoner and exile; it allows the subject of feminism to occupy both the center and margins, the inside and the outside." Building from the work of others, G. Rose (1993) argues that recognizing our doubled positions (both inside and outside) can be politically productive because it both allows us to problematize and exhaust the meanings of margin and center and sharpens our critical capacity. By understanding that we as

19

individuals move between/across margins and centers, we can destabilize unexamined dualisms and boundaries as we begin to see the inherent connections between inside/outside, center/margins, same/other.

The attractions of mobility are perhaps most systematically articulated by Kathy Ferguson's (1993) use of the metaphor of mobile subjectivities; this metaphor accentuates movement without defining what it is that subjects move between, to, or from. Ferguson also presents the metaphor as a strategy to disrupt dualistic thinking and essentializing around any social categories: "I have chosen the term *mobile* rather than *multiple* to avoid the implication of movement from one to another stable resting place, and instead to problematize the contours of the resting one does" (158). "Class, like race, gender, erotic identity, 'etc.,' can be a crucial but still temporary and shifting resting place for subjects always in motion and in relation" (177).

Spivak also speaks of the attractions of movement, in her case, actual geographical as well as subjective mobility: "As far as I can tell," she says "one is always on the run, and it seems I haven't really had a home base – and this may have been good for me. I think it's important for people not to feel rooted in one place. So wherever I am, I feel I'm on the run in some way" (1990: 37). Many feminists have narrated a process of uncovering multiple layers of social identity by departing from their home place. Through traveling to Cuba from the United States, for example, Johnetta Cole (Bateson 1990: 45) recalls how she began to understand herself in gendered and not just racialized terms: "There I was," she says, "seeing for the first time the possibility that the race thing was not forever and ever; and then the other -ism [sexism] was right up there saying, what about me?" Teresa de Lauretis (1988: 128) tells of the importance of immigrating to the United States for her awareness of ethnic difference: "[My] first (geographical) dis-placement [from Italy to the United States]," she writes, "served as a point of identification for my first experience of cultural difference (difference not as simple distinction, but as hierarchized)."

Some feminist writers fold contemporary notions of decolonized consciousness and situated networks of feminist politics onto their understanding of places. Chandra Talpade Mohanty (1987: 41), for example, counters the notion of space as territory, which she links to "the logic of imperialism and the logic of modernity," with an understanding that:

[i]n North America of the 1980s geography seems more and more like "an abstract line that marks the separation of the earth and sky." Even the boundaries between Space and Outer Space are not binding any more. In this expansive continent, how does one locate

oneself? And what does location as I have inherited it have to do with self-conscious, strategic location as I choose it now?

(Mohanty 1987)

Mohanty leaves these as open questions.

Drawing on Virilio's (1986) speculations about the impact of information technology on urban form, Elizabeth Grosz (1992) seems to answer these questions by severing the links between subjectivity (and the body) and space as bounded territories: "The subject's body will no longer be disjointedly connected to random others and objects according to the city's spatio-temporal layout. The city network – now vertical more than horizontal in layout – will be modeled on and ordered by telecommunications" (252). (See also Emberly 1989; Robins 1991 for nonfeminist variants of this argument.)

We interpret our case study of Worcester as tempering some of the excesses of mobility stories conceived in frictionless space. Taken to an extreme, claims about mobile subjectivities could lead us to underthematize the "resting places" of identities and the institutional apparatus that fixes identities (Pratt 1993). Just as Janet Wolff (1993) has argued in reference to travel metaphors in cultural theory, there is a danger that they will lull us into forgetting that relations of domination persist and continue to fix us in social and geographical spaces. One part of the process of disrupting boundaries involves a careful and continuous examination of how they are constructed and reconstructed.

Contemporary claims about post-territorial or post-Euclidean space also seem like oversimplifications and, taken to an extreme, obscure one of the media through which identities get fixed, namely through places. As Massey (1992) has argued, "[a] special type of hype and hyperbole has been developed to write of these matters. The same words and phrases recur; the author gets carried away in the reeling vision of hyperspace" (8). Among other problems, Massey notes that this oversimplified reading of contemporary spatial relations fails to register that different social groups, "and different individuals belonging to numbers of social groups, are located in many different ways in the new organization of relations over time–space. . . . [M]uch, if not all, of what has been written has seen this new world from the point of view of a (relative) elite" (9).

What seems particularly interesting (and baffling) about places is the complex layering of boundedness and fluidity, localism and globalism (and Massey argues that, in principle, there is nothing new about this). Again in Massey's words: "Even as you wait, in a bus shelter in Harlesden or West Brom, for a bus that never comes, your shopping bag is likely to contain at least some products of the global raiding party which is constantly conducted to supply consumer demands of the

21

world's relatively comfortably-off" (9). Attention to the globalization of culture and economy should not allow us to lose sight of the rootedness of local lives, to facts of existence such as the tedium of waiting for buses and the effects that this waiting might have on the rest of one's life.

In our study of Worcester, we are especially attentive to the resting places of identity (especially class and gender ones) and to the construction of territories (labor markets and social worlds) within cities, and to how the resting places of identity are bound up with the territories in which people live. We are not suggesting that local places are sealed off from other places by fixed or static boundaries. But we are arguing that social and economic boundaries still exist within Worcester and that individuals are connected to others according to the city's spatio-temporal layout in ways that make a difference to their experiences of gender, class, and race, among other social relations. In our case study we try to understand how contemporary social and economic boundaries are constructed, in part out of previous ones, and the numerous ways that local places and identity intersect, overlap, and shape each other. We see our study as tempering and complementing contemporary mobility stories.

Situated theories

In his description of the distinction between the experience of space and representations on a map, de Certeau (1984: 98) reproduces a typically modernist account of the flâneur, moving freely through urban space: "the walker transforms each spatial signifier into something else. And if on the one hand he actualizes only a few of the possibilities fixed by the constructed order (he goes only here and not there), on the other he increases the number of possibilities (for example, by creating shortcuts and detours)." He creates "a rhetoric of walking" that "cannot be reduced to the graphic trail" (99). The gendering of this description is entirely right; as numerous feminists have observed (Valentine 1989; Pollock 1988; Wolff 1985, 1990, 1993), the flâneuse is a practical impossibility: legitimate fears about safety and societal prohibitions restrict women's walking possibilities, creating a different, feminine, "rhetoric" of walking. (Placed against this, we should also note that some feminists acknowledge the liberatory potential offered by the social diversity and anonymity of urban life: Wilson 1991.)

Though de Certeau seems oblivious to the partiality of his account, feminists have long argued that theories are the products of embodied subjects, situated in the contingency and partiality of the theorist's experience. What distinguishes the position of many feminists, however, is a simultaneous commitment to this radical social constructivist perspective and a belief that empirical research – studies of "the real" –

can be used to construct a better, more equitable world (Haraway 1991). In an effort to hold on to both ends of this dichotomy, Haraway attempts to ground a notion of objectivity in the positionality of the theorist. This is a perspective that she interprets as being different from and even hostile to relativism insofar as the latter fails to ground theory in the situation of the theorist. Relativism is "a way of being nowhere while claiming to be everywhere equally" (Haraway 1991: 191). In contrast, she argues that situated knowledge claims are based on the recognition that all knowledge is interpreted *from a location*. That location and its effect on vision (or interpretation) then become a serious matter for scrutiny. It is also a beginning point for carefully building webs of connection with others, who partially share our locations and who then build bridges with others who overlap in their locations in different ways.

Situated knowledge is thus grounded in a critical examination of theorists' locations. This concern with the "politics of location" (Rich 1986) flows through and structures much of the debate in feminist studies since the mid-1980s. It has two very important repercussions for our empirical work in Worcester: first, it forces a critical rethinking of the central analytical category that has organized our thought, that of gender, and, second, it presses us to position ourselves and to reflect on how our own situations structure our interpretations.

A critical appraisal of location has been forced by women who argue that white, middle-class, heterosexual, Western academic feminists have been guilty of their own universalizing gestures; we have not been sensitive to the particularity of our situations and the partiality of our knowledge claims. This has spawned a burgeoning literature in which African American women, Latinas, lesbian women, women from "developing" countries, working-class women, and many others, speak of and from their worlds and their experiences.

A recognition of differences among women slips into debates about essentialism and the analytical and political status of the category "woman." There seem to be at least two meanings of essentialism within the feminist literature. One grounds women's common experience in the female body. A less empirical interpretation of essentialism refers to the speaking of women as a category of beings.

There is a sense within some recent feminist writing that the debate about essentialism has become increasingly sterile, that various meanings of essentialism are confused (de Lauretis 1989; Fuss 1989) and that the label is often used simply to silence. "What revisionism, not to say essentialism, was to Marxism-Leninism, essentialism is to feminism: the prime idiom of intellectual terrorism and the privileged instrument of political orthodoxy . . . the word essentialism has been endowed within the context of feminism with the power to reduce to silence, to excommunicate, to consign to oblivion" (Schor 1989: 40). This has provoked some critical rethinking of the socially constructed relations between femininity and the body (Grosz 1992; Schor 1989), as well as the conditions under which it is appropriate to use the category "woman."

A first step toward clarity seems to rest on the recognition that one can speak of women as a group without submitting to the idea that it is nature that binds the category (i.e., to recognize the distinction between empirical and nominal essentialism) (Fuss 1989). Ellen Rooney (in Spivak 1993: 2) likens the desire to ground feminism and the category "woman" in the empirical, in the biological body, to "the old dream of 'non-partisanship' at the heart of politics [in the United States] . . . a dream of the end of politics among women." Instead of rejecting politics, feminists who reopen serious consideration of nominal essentialism ground it *in politics*, in a strategic political choice to conceive of women or groups of women as a category (Fuss 1989; Spivak 1993). Whether the strategy has liberatory or revisionist effects depends on the situation: on who uses the strategy, on the audience, on specifics of the historical situation; and it is a strategy that must always be held in tension with a critical impulse that continually forces to the forefront political and strategic questions such as: "what regulates your diagnosis, why do you want me with you, what claims me, what is claiming me?" (Spivak 1993: 21).

The "choice" to essentialize is not an entirely free one: in a patriarchal culture that constructs embodied subjects in sexual and gendered terms, there is little choice but to work with these categories, recognizing that they are constructed ones. We filter our investigations through the categories of gender in an effort to understand how and why women continue to be constructed as workers in the secondary sectors of the labor market. We explore the discursive and material conditions that lead women to position themselves in gendered terms and look at the gendered and spatial codes that employers work with, ones that lead to the gendered and spatial segmentation of local labor markets. At the same time, we recognize that all women are not situated within secondary labor markets and not all of those who are thus situated are positioned within secondary labor markets in the same way. Discussions of differences among women have sensitized us to the necessity of "pluralizing the grid" through which we filter women's experiences, to recognize racial, ethnic, and class differences, and differences in family circumstances and individual women's lives across the life course.

This grid must be place-specific. Much of the feminist debate about essentialism takes place in the abstract, a factor that may account for its increasing sterility, and one that is strikingly at odds with the emphasis on situatedness, of bodies, authors and theories. If one takes Spivak's (1993)

24

comments about strategic essentialism seriously, it makes little sense to speak about difference and essentialism out of contexts. Various grids of difference emerge in different places, and the strategy of essentialism must be evaluated in place. Spatial stories such as these have not been carefully explored in the feminist literature. One of our objectives is to examine the spatial stories and experiences that are part of the process through which differences are constructed in Worcester. This stands as an attempt to bring another perspective to feminist discussions of essentialism and difference, by tracing the spatiality of the construction of differences (see also Hanson 1992; Pratt and Hanson 1994).

This perspective reflects our disciplinary home. If we absorb Haraway's position on situated knowledge as a feminist route toward objectivity, we should try to think more deeply about our own positions and the ways that they structure our vision. At this point, we just comment on the partiality of our vision; this is a theme that we return to in Chapter 3, when we outline our research strategies more clearly.

Positions are not static; this is a point that needs to be underlined carefully in the contemporary context, in which "marking" by sexual orientation, class, race, etc. is sometimes used not only to open up new conceptual spaces but also to discipline and silence others. Encounters with different literatures have altered our perspective by showing us some of our own blindspots. The discussion in this chapter has outlined our travel itinerary over the last six years and isolated some of the ways that our vantage points have shifted.

A MAP OF THINGS TO COME

The next chapter provides a contextual backcloth for subsequent ones, by providing a historical overview of the social and economic development of Worcester over the last century. Of course Worcester is at once typical and distinctive. By placing Worcester in and against its New England and American context, we try to tease out the ways in which Worcester contributes to, and at the same time deviates from, national patterns. In Chapter 3 we describe our research strategies, first in a straightforward and then in a self-reflexive way.

In Chapter 4 we describe the gendered spatial patterns that characterize the geography of employment. Like women elsewhere, women in Worcester experience the friction of distance differently from men; moreover, the geographical mobility of women who work in female-dominated occupations is especially constrained, so that occupational segregation has a geographic dimension. The map of employment in Worcester also bears the stamp of gender, with the gender composition of jobs varying substantially from one small area to another. We consider the impact of the spatial distribution of employment opportunities on occupational segregation by asking whether a woman's residential location on the variegated employment map affects the type of work she has.

In Chapter 5 we look inside households in Worcester to understand how the various household arrangements affect women's labor market situations and help to create the patterns described in Chapter 4. We tell fairly conventional geographical stories about these arrangements: namely, those women with the heaviest domestic responsibilities tend to work closer to home and be most disadvantaged in the labor market. We attempt not only to impress upon the reader the sheer number of arrangements that lead to this conclusion, but to acknowledge deviations from the norm, in an effort to disengage our analysis of labor market situations from uncritically essentialized concepts of "woman" and "man."

In Chapter 6 our attention shifts to the construction of very local labor markets and, in particular, the role that employers play in this process, often through their attempts to gain access to "pools" of labor with specific social characteristics (including gender ones) but also in employers' roles in shaping those labor pools in distinctive ways. In this discussion we begin to explore how the friction of distance not only reflects and reproduces gender divisions of labor, but enters into the constitution of places. Chapter 7 complements the previous one by showing how employees act in concert with employers to create local labor markets. We also attempt to trace out some of the implications of local intraurban labor markets by considering how different gender and class identities are constructed in four areas within Worcester. In other words, we trace another step in the process, by exploring how places, in part constituted through gendered home–employment distances, play a part in constructing distinctive gendered and class experiences and identities.

Our spatial stories move, then, between a preoccupation with distance, to a focus on place. It is not simply that different gender and class identities emerge in different places, but that they are constituted in those places, in part because of the different histories and reputations of those areas. An understanding of how boundaries – around labor markets, neighborhoods, social networks, and specific gendered household arrangements – are constructed, we hope will allow women and men in Worcester (and elsewhere) to move through and across them.

2

CONTINUITY AND CHANGE IN
WORCESTER, MASSACHUSETTS

Woman is no longer confined to a narrow sphere; the whole world is
hers.

> (Georgia Bacon, President of the Worcester Woman's
> Club, speaking to the Worcester Society of
> Antiquity in 1900 (quoted in Feingold 1986: 13))

In 1913, the Research Department of the Women's Educational and
Industrial Union of Boston published a pamphlet entitled *A Trade School
for Girls: Preliminary Investigation in a Typical Manufacturing City, Worcester,
Mass.* The pamphlet reported on a study that had been designed to
explore the need for and feasibility of establishing a trade school for girls
in Worcester (members of the business elite had started a Boys' Trade
School in 1910 to prepare boys for work in Worcester's metal and
abrasives industries).

In the spring and summer of 1911 the Boston researchers visited 214
Worcester families in their homes to talk with parents and their
daughters about school and paid work; they also interviewed employers
in 63 Worcester firms about their employment practices. The study was
prompted by concern over:

> the great army of young girls who go out to employment as soon as
> they have passed beyond the reach of the compulsory law; the
> number of girls and women who are employed in undesirable
> industries; the lack of opportunity for advancement and better wage
> earning which confronts the average female wage worker; . . . the
> instability of female as well as male workers in many industries; the
> fluctuating character of their employment and the low wage which
> most of them are able to earn.
>
> (Research Department of the Women's Educational
> and Industrial Union of Boston 1913: 6)

The women-employing industries of Worcester at the time were
textiles, women's and children's apparel (corsets, underwear, women's

wear), wire and metal goods, and envelopes and paper products. Within these industries most women and girls were employed as machine operators in jobs that were repetitive and monotonous, offered low pay with no chance for learning or advancement, and had high rates of turnover. Seeing a connection between the high rates of worker turnover and the mind-numbing working conditions, the researchers noted somewhat dryly that while they found employers to be concerned about high turnover, "none has yet attempted to solve the problem through systematic training of their workers" (13). Instead, the high job turnover among girls had prompted employers at "the better factories" (30) simply to stop hiring them.

Despite these grim features of paid employment, girls were provoking concern by leaving school in unprecedented numbers, usually when they were 14 years old. The authors of *A Trade School for Girls* debunked a popular myth of the day when they discovered that fully half of these girls left school and sought work voluntarily, not out of economic necessity. Contrary to prevailing thought, which held that girls entered the labor force solely at their parents' behest in order to contribute to the household's sustenance, the researchers learned that half of the girls who were working in factories were doing so because they found paid work more

Plate 2.1 This 1910 photograph of Royal Worcester Corset company employees reveals a largely female labor force. Photograph from the Worcester Historical Museum, reprinted with permission

attractive than continuing in school. (The researchers explain this by noting that children, including girls, are either "book-minded" or "motor-minded"; it is the "motor-minded" girls who, disenchanted with school, are likely to leave it for paid employment, and it is for the "motor-minded" girls that educational opportunities in the form of a trade school should be established.)

Certain jobs, especially those considered most "skilled," were closed to women, usually by dint of men having organized to block women's access to their jobs. For example, in clothing manufacturing the well-paid job of cutting was monopolized by men, and wireworkers and wool spinners had struck to keep their jobs all-male (Rosenzweig 1983: 18). In some cases these prohibitions against the gender integration of jobs were backed by state law; for example, women were excluded by state law from the core rooms of foundries, and the 56-hour law, prohibiting women from working in factories or stores more than 56 hours per week, excluded women from most weaving jobs (Feingold 1986). The extent of early twentieth century sex segregation in the Worcester labor market is visible in the 1910 photograph of the Royal Worcester Corset Company's employees (most of whom were women), assembled in the factory's courtyard (Plate 2.1).

The Boston researchers conclude their 1913 study with a realistic assessment of women's opportunities in the Worcester labor market and an evaluation of how women and girls might optimize their position within that market. Fully cognizant of the tenacious hold of occupational segregation by gender (and not expecting to loosen it), the authors ask which "women's jobs" require the greatest skill (and therefore command the highest pay) *and* are currently in greatest demand in Worcester *and* currently offer little or no on-the-job training; these are the jobs that a Worcester Trade School for Girls should target.[1] Their verdict: machine operating, dressmaking, and millinery.

We recount in such detail the findings reported in *A Trade School for Girls* because we are struck by how remarkably little has changed in the 75 years between the early 1910s, when the Research Department of the Women's Educational and Industrial Union of Boston studied women's employment in Worcester, and the late 1980s, when we did. True, state law no longer bars women from certain jobs. But the segregation of jobs according to gender, the low status and wages accorded to women's work, the lack of opportunity for advancement or on-the-job training, the role of employer discrimination in shaping job opportunities – all faced Worcester (and American) women seeking employment in the late 1980s as they did in the early 1900s. Furthermore, now, as then, women work for a variety of reasons, including the need to contribute to household income as well as the desire for challenge or personal fulfillment.

The 1913 study even comments on a couple of geographical dimensions, which our recent study confirmed as enduring. The authors of *A Trade School for Girls* report that in 1913 Worcester was "a political, industrial, and social entity, resulting in a lack of interchange of work and workers with Boston" (p. 13). We found that, despite the dramatic improvements in transportation since 1913, Worcester remains a relatively self-contained labor market, quite separate from Boston's (see Figure 2.1).[2] The authors of that earlier study also recognize (27) the importance of both the job opportunities that are within walking distance of home and the occupation of relatives (especially mothers and sisters) in determining the type of work a girl takes up. As we describe in detail in Chapter 4, the nature of jobs that are located close to home remains of crucial importance to working women, and as we note in Chapter 7, the occupations of family members influence women's job choices. Surely the 1913 study supports Georgia Bacon's claim that in 1900 women's lives were "no longer confined to a narrow [domestic] sphere": many of Worcester's factories were running on largely female labor. But just as surely the "whole world" was *not* "hers," unless one took Worcester, or some small part of it, to be the whole world, or unless

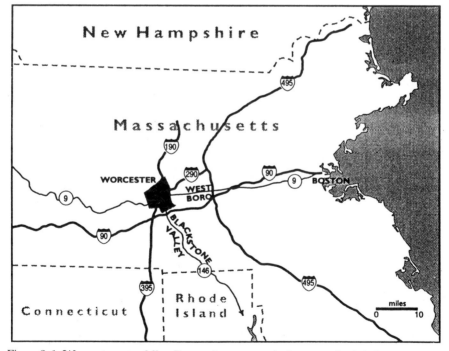

Figure 2.1 Worcester straddles Route 9, a route dating to colonial times, but is poorly connected to Interstate 90 (the Massachusetts Turnpike), constructed in the 1950s (see note 2)

one saw employment in a narrow range of poorly paid jobs as opening up global horizons.

As they signaled in the title, the authors of *A Trade School for Girls* saw Worcester as a typical manufacturing city and chose to study it in part for its representativeness. At the same time, they stress "the necessity of knowing local conditions" (9) before drawing conclusions, and their analysis and recommendations are certainly tuned to the distinctive resonances of the Worcester labor market of the time. In our investigation of contemporary Worcester, we, too, have bifocal vision, appreciating both what is general (or typical) and what is particular (or peculiar) about this place. We chose Worcester neither for its likely typicality *nor* its distinctiveness; rather, Worcester became the laboratory for testing our ideas and for learning about gender, work, and space because it is a place we are both familiar with. It is also a place that is at once large enough to be truly urban and small enough to be manageable.

It is in many ways fitting that we as feminists would serendipitously settle on Worcester as the locale for our study, for it is a city that, prior to rapid industrialization in the late nineteenth century, displayed significant progressive impulses, including flashes of feminism. In 1830, thanks largely to the efforts of Worcester's Dorothea Dix, Worcester became the site of the country's first state-supported institution for mentally ill people, and at mid-century the city established the country's first municipal park. Oread Collegiate Institute opened in 1849 (and lasted until 1881) as the nation's first women's college offering the same four-year liberal arts curriculum previously available only to men.

Suffragists were active in Worcester and organized the first national Women's Rights Convention, which was held in the city in October of 1850; a year later the second national Women's Rights Convention was also held in Worcester. Attended by more than one thousand people (and, according to Elizabeth Cady Stanton, Susan B. Anthony, and Matilda Joslyn Gage (1881: 225), with attendance limited only by the size of the convention hall), these conventions drew the leading suffragists and abolitionists of the time. Lucy Stone, Frederick Douglass, Abby Foster Kelley, Sojourner Truth, Lucretia Mott, and William Henry Channing – all came to Worcester to demand equal political and economic rights for women. In their *History of Woman Suffrage* (1881), Stanton, Anthony, and Gage reflect on these mid-century meetings in Worcester: "It may be said with truth, that in the whole history of the woman suffrage movement there never was at one time more able and eloquent men and women on our platform, and represented by letter there, than in these Worcester Conventions" (242).[3]

But the same feminist feistiness that brought women activists to Worcester in 1850 was blamed in a 1981 publication (Erskine 1981) for the downfall of one of Worcester's leading industries, women's corsets

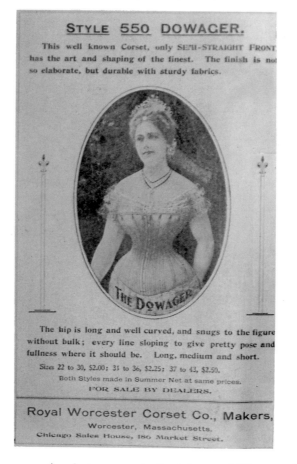

Plate 2.2 The connection between early feminism and Worcester's industrial fortunes were drawn in the caption beneath this Royal Worcester Corset Co. advertisement for the "the dowager" that appeared in Erskine (1981: 88). The caption read: "During the Women's Suffrage Convention held in Worcester in 1850 women complained that they would never be free until they changed their cumbersome mode of dress. After World War I, women bobbed their hair, shortened their skirts, and threw away their corsets. The changing styles brought an end to one of Worcester's most prosperous industries, corset manufacturing." Photograph from the Worcester Historical Museum, reprinted with permission

(see Plate 2.2). A national center of wire production since the 1830s, Worcester developed corset manufacturing as a related, woman-employing industry that spun off from and "complemented" the male-dominated production of wire.

Erskine's caption (1981: 88) for the historic illustration advertising "the dowager" corset (Plate 2.2) signals both the reasons for, and the

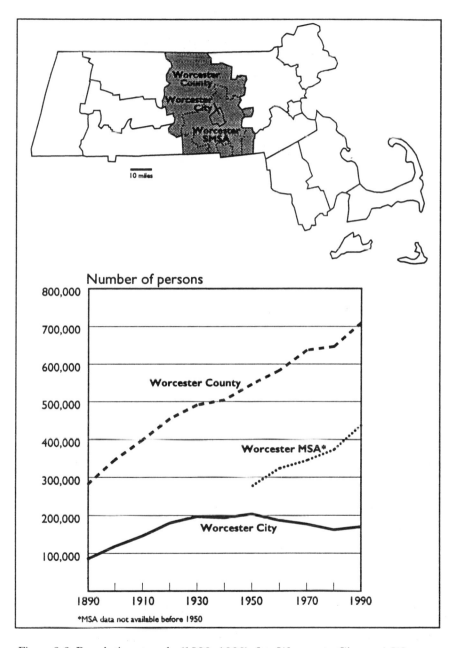

Figure 2.2 Population trends (1890–1990) for Worcester City and Worcester
County and for the Worcester Metropolitan Area (1950–1990)
Source: U.S. Census Office 1890; 1900; 1910; U.S. Bureau of the Census 1920; 1930;
1940; 1950; 1960; 1970; 1980; 1990a

complexities of, focusing on "women's employment." On the one hand, men neither sewed corsets nor wore them; women were clearly the producers and consumers. On the other hand, how many of Worcester's female corset stitchers in the 1910s (and there were nearly 1,200 of them, outstripping the number of women in any other kind of work (Rosenzweig 1983: 13)) were among those who wore corsets and later discarded them in the 1920s? To talk of "women's employment" is necessarily to highlight and understand important – and very real – gender differences, but it can submerge important differences of, *inter alia*, class, ethnicity, and age among women.

In the remainder of this chapter we offer glimpses of Worcester by way of providing historical depth and context for our contemporary study. The discussion is loosely sorted into sections on population trends and employment patterns, the impacts of immigration, Worcester's anti-union ethos, and local political culture.

POPULATION AND EMPLOYMENT TRENDS

The Worcester that the Women's Union of Boston studied in 1911 was a thriving industrial city of 145,000 people, the third largest city in New England and twenty-ninth largest in the U.S. The number of people living within the city limits has fluctuated over time, peaking at more than 200,000 in 1950; the lure of the suburbs then steadily drained population from the city, until the 1980s, when a lucrative housing market prompted residential developments on still-open land within the city (see Figure 2.2). Meanwhile, the population of the metropolitan area has grown apace since 1950, when the U.S. Bureau of the Census first carved out Standard Metropolitan Statistical Areas (SMSAs) (by the 1990 census these had been changed to MSAs by dropping the "Standard"). Within Massachusetts, Worcester is the third largest metropolitan area, after Boston and Springfield.

Throughout these population swings, Worcester has never been a one-industry town. The diversity of industries employing women in 1913 – corsets, textiles, wire, and envelopes – is indicative of the industrial diversity that has been a hallmark of the Worcester economy since at least the mid-nineteenth century. Even in 1865 when Worcester's leading industrial activity was the making of boots and shoes, less than one-quarter of the labor force was employed in that industry (Kolesar 1989: 8). In contrast to other Massachusetts manufacturing towns like Fall River and Lynn, where more than four-fifths of the work force labored in one industry (cotton mills and shoes, respectively), Worcester throughout the late nineteenth and early twentieth centuries never had an industry that employed more than one-quarter of its workers (Rosenzweig 1983: 12).

34

Percentage of employment in manufacturing and services

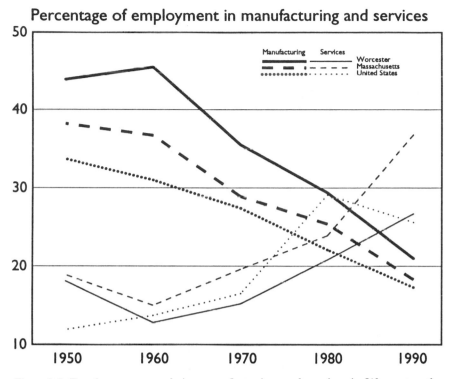

Figure 2.3 Employment trends in manufacturing and services in Worcester, the state of Massachusetts, and the U.S., 1950–90

Source: Massachusetts Department of Employment and Training 1975; U.S. Bureau of the Census 1950; 1960; 1970; 1980; 1990a

While no single industry or firm has dominated the city's growth, manufacturing has been crucial to the Worcester economy, employing a higher proportion of the local labor force than the U.S. average ever since at least 1870.[4] Industries sprouted and thrived in Worcester after the 1828 opening of the Blackstone Canal linking Worcester to Providence, Rhode Island, after railroad connections linked the city to raw materials and markets in the late 1830s, and after the advent of steam power in 1840 reduced the reliance on water as a power source (Rosenzweig 1983: 11). In addition to the "woman-employing" industries already mentioned, traditional Worcester industries have included abrasives and ceramics, jet engines and automobile crankshafts, tool and die manufacturing, and heavy machinery. Since the 1960s, these industrial stalwarts have been joined or replaced by contemporary high-technology firms (manufacturing computer hardware and software, optics, and biotechnology products) as well as by some high-technology spinoffs from long-term Worcester specialties like ceramics, abrasives, and wire.[5]

35

Worcester's industrial diversity long insulated the local economy from the fluctuations buffeting single-industry cities, but it has not been able to stave off the fundamental sea-change transforming the U.S. economy since the 1960s: the decline in manufacturing and the concomitant rise in service employment. Figure 2.3 documents, for Massachusetts and the Worcester MSA, the precipitous erosion of jobs in manufacturing since 1960 and the steady accretion of service-sector jobs. Though badly battered by 1990, Worcester's manufacturing sector nevertheless still claimed a higher segment of the metropolitan-area labor force (21 percent) than did manufacturing in Massachusetts (18.3 percent of the state's work force) or the U.S. (17.3 percent). Services, employing 26.7 percent of Worcester-area workers in 1990, and finance, insurance, and real estate (FIRE), employing 7.3 percent, also both exceeded national averages (25.6 percent and 6.2 percent respectively).

Worcester's past reliance on manufacturing – like that of Massachusetts – makes the city vulnerable to job loss, as manufacturing firms seek lower-cost sites or simply cease operations altogether. During the 1980s Massachusetts suffered a larger decline in manufacturing jobs than any other state (*Forbes* 1988), and between 1988 and 1992, Worcester (city) lost nearly 17 percent of all its jobs (Stein 1993). The Worcester area has been successful in attracting several large biotechnology firms in an effort to replace some of the lost manufacturing jobs. The fact that industrial rents are 30 percent lower in Worcester than in nearby Cambridge, Massachusetts, a major competitor in the scramble to lure the biotechnology industry, is a substantial inducement to locate in Worcester. Also attractive are Worcester's housing prices, which are significantly lower than the cost of comparable housing located closer to Boston.

Worcester's central location makes the city an entrepôt for the warehousing and distribution of goods. Six million people (including those living in Boston and Springfield, Massachusetts and Providence, Rhode Island) live within 50 miles of Worcester, and in 1987 the U.S. customs service designated this landlocked location a "port city." Worcester now is a site for receiving, breaking down, and shipping containerized goods off-loaded from ships at Boston and Providence.

Worcester has been home to several large insurance companies since the mid-nineteenth century and to several colleges and universities since the late-nineteenth century, so that the city's economic diversity has long included a healthy dose of employment in producer services (Plate 2.3). That sector of the economy received a major boost when, in the early 1970s, the state launched the University of Massachusetts Medical School and Hospital in Worcester; since then, the UMass Medical Center has grown to become Worcester's second largest employer, surpassed only by the City of Worcester (City of Worcester 1991). Just as Worcester

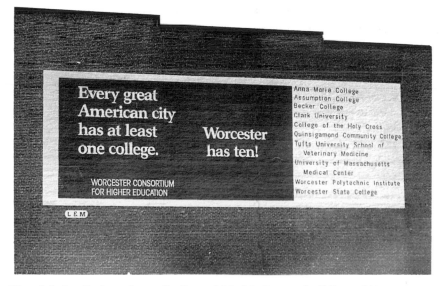

Plate 2.3 Applied to the wall of an old brick factory building, this announcement, visible to motorists on the thruway that runs through the city (Interstate-290), signals the increasing importance of higher education (a total of some 17,000 full-time students) to the Worcester economy. Photograph by John Ferrarone

Polytechnic Institute (WPI) has, since 1865 nurtured the engineering skills that have powered Worcester's traditional manufacturing sector, so now does the UMass Medical Center incubate the scientific base on which Worcester's nascent biotechnology industry is growing.[6] These place-based synergies between the manufacturing and service sectors have a long tradition in Worcester.

We undertook our study of Worcester at the end of the 1980s, when both the Worcester and the Massachusetts economies were still booming thanks to the "Massachusetts Miracle." Between 1950 and 1979, New England and Massachusetts lagged behind the U.S. in per capita income growth and rates of job creation and had unemployment rates that exceeded the U.S. average (Barff 1990). Between 1979 and 1986, New England fared better than the U.S. on all these indicators (Massachusetts Institute for Social and Economic Research 1994). This economic turnaround can be traced to the manufacturing sector, although only one-sixth of the new jobs added in Massachusetts between 1979 and 1986 were in high-technology industry (Harrison 1988), and many of these were related to federal military spending. During the 1980s Massachusetts benefited disproportionately from federal defense spending, reaping, on a per capita basis, roughly three times the national average between 1987 and 1989, for example, in federal defense dollars

(Henderson 1990: 6). Browne (1988) has outlined the strong links between federal defense spending and high-technology development in Massachusetts; and Harrison (1988) has estimated that one-quarter of the new jobs in high-technology manufacturing in Massachusetts during the 1980s were dependent on military spending.

The economic boom did not exempt Massachusetts or Worcester from the increasingly bifurcated wage structure that accompanies the shift from manufacturing to service jobs (Bluestone and Harrison 1982; Herwitz 1987). In 1985, the Worcester area lost 2,486 manufacturing jobs and gained two-and-a-half times that number (6,389) in service sector employment. The average manufacturing wage that year was $23,200, about 30 percent more than the average service worker's wage ($17,800) (Herwitz 1987).

As many have noted (e.g., Moscovitch 1990; Case 1992), the rapid economic growth of the 1980s sowed the seeds of its own destruction. Most of the job growth associated with the "Massachusetts Miracle" was confined to the Boston area (Harrison 1988); the boom led to skyrocketing prices in the Boston housing market, hampering the recruitment efforts of Boston-area firms. By 1987 the median price of a single family home in the Boston area was more than twice that for the U.S. and had increased 155 percent since 1983 (Case 1992). The wave of rising real estate prices rolled westward, washing over Worcester in the mid-1980s. The average price of a single-family home in the Worcester area jumped from $56,500 in 1980 to $165,400 in 1988 (Greater Worcester Multiple Listing Service, Inc. 1994). By 1990 the average had dropped to $153,400 and had continued falling to $141,900 in 1992.

Throughout the prosperous 1980s Massachusetts continued to experience net outmigration, though at a rate much reduced from the previous three decades (Barff 1990; Sege 1989). At the same time, the state had a low rate of natural population increase, ranging from 3.0 percent in 1980 to 6.4 percent in 1989 (Commonwealth of Massachusetts, Department of Public Health 1992) which, together with the net outmigration, contributed substantially to the remarkably low unemployment rates in the mid- and late-1980s (Harrison 1988), precipitated labor shortages, and fed rising wages. Whatever "miracle" had visited the Massachusetts and Worcester economies in the mid-1980s vanished abruptly at the end of the decade; by 1990, the unemployment rate was on the rise in Massachusetts and in Worcester, again exceeding the national average and doing so in Worcester for the first time in 15 years (Figure 2.4).

The boom of the 1980s with its attendant labor shortages drew many women and teenagers into the paid labor force; by 1989 New England had the highest female labor force participation rate (60 percent) of any region in the country – also the highest teenage participation rate (56 percent) (Sege 1989). Much of the "Miracle's" prosperity actually came,

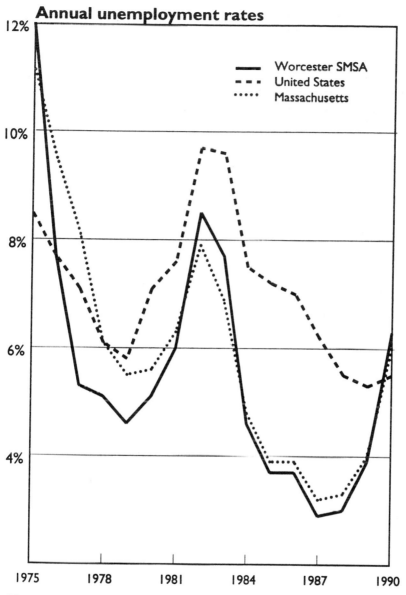

Annual unemployment rates

Worcester SMSA
United States
Massachusetts

Figure 2.4 Annual unemployment rates, Worcester MSA, the state of Massachusetts, and the U.S., 1975–90
Source: Anon. 1975; U.S. Bureau of the Census 1980; 1990a

therefore, from increases in the number of workers per household, as well as from longer hours of work. But women have always been a sizable portion of the Worcester-area labor force and, as we have seen, a significant presence in particular industries. As early as 1837, females made up 58 percent of workers in Worcester woolen mills, 38 percent in the hat firms, 27 percent in the shoe and boot factories, and 50 percent in the paper mills (Feingold 1986).[7] Hill (1929) reported that of 63,454 women (aged 16 and over) in Worcester in 1920, 20,434 were gainfully employed, yielding a labor force participation rate of more than 32 percent.

The proportion of Worcester's working-aged women in the paid labor force grew from about one-quarter in 1890 to about 30 percent in 1910. It remained roughly at that level until the 1950s when the proportion again began rising, exceeding half of women aged 16 to 64 in 1990 (see Figure 2.5). Whereas prior to 1970, women living in the city were more likely than were suburban women to have a paying job, since that time

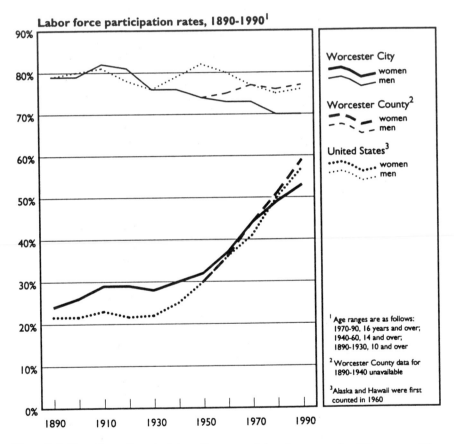

Figure 2.5 Female and male labor force participation rates for Worcester City, Worcester County, and the U.S., 1890–1990

Source: U.S. Census Office 1890; 1900; 1910; U.S. Bureau of the Census 1920; 1930; 1940; 1950; 1960; 1970; 1980; 1990b

the tables have turned, with labor force participation rates for suburban women exceeding those of their city-based sisters. This shift is doubtless connected to the suburbanization of employment, which brought jobs closer to suburban residential areas. Since 1950, when data on participation rates for Worcester County first became available, the percentage of

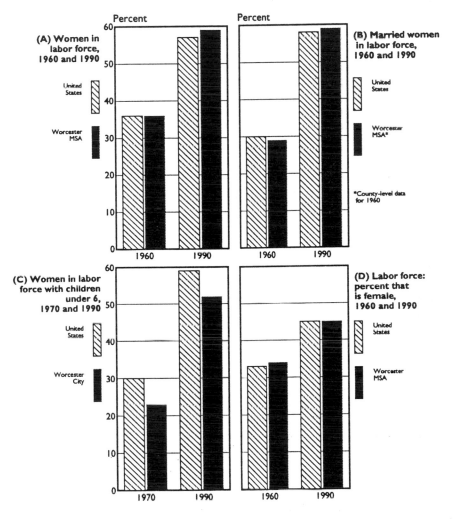

Figure 2.6 Changes over time in female labor force participation, Worcester and the U.S. (A) labor force participation of women, 1960 and 1990; (B) labor force participation of married women, 1960 and 1990; (C) labor force participation of women with children under age 6, 1970 and 1990; (D) proportion of labor force that is female, 1960 and 1990

Source: U.S. Bureau of the Census 1960; 1970; 1990b

women in the labor force in Worcester County has departed little from the percentage in the U.S.

In Worcester, as in the U.S., participation rates have risen most dramatically among married women and women with preschool children, with the proportion of married women in the labor force growing from less than one-third in 1960 to well over one-half in 1990 and the proportion of working women with young children increasing from 23 percent of such women in Worcester in 1970 to 52 percent in 1990 (Figure 2.6). In both of these measures, as in female labor force participation rates and in the proportion of the labor force that is female, trends in Worcester closely match those for the U.S.[8]

Smith notes that the increase in female labor force participation between 1960 and 1982 in the United States occurred disproportionately among single and married women (i.e., not among divorced, separated, and widowed women), leading her to conclude that:

> The great increase in the participation of married women in the wage labor force was a response by white working-class families to the decline in the value of their wage. The wages of all working-class wives, whether white or black, became the crucial ingredient in keeping families out of poverty.
>
> (Smith 1987: 424–5)

Worcester's American context means that women's participation in waged work takes place in an environment largely devoid of the state-provided social services – especially health care, child and elder care – available in other countries. Child care, particularly child care of a quality acceptable to parents, is in short supply relative to the demand. Formal child care (as opposed to that provided by friends or relatives) is a market good, which is often very costly. In Massachusetts, women who receive Aid to Family with Dependent Children (AFDC), a state-provided welfare payment, are eligible, when they go to work outside the home, also to receive state funding to defray the cost of day care, which is provided through privately run centers. Because so few state-subsidized day care places are available, many (and in some areas, most) low-income working women have to manage with unsubsidized day care, the high cost of which forces some women out of the paid labor force and back into the home.

Moreover, until 1993, many workers (usually women) were fired from their jobs when pregnancy, childbirth, adoption, or the illness of a family member meant they had to leave the labor market temporarily. Under the 1993 Family and Medical Leave Act, employers with fifty or more workers are now obligated to provide up to twelve weeks of unpaid leave for medical emergencies, childbirth, or adoption. Two-thirds of the work force are covered by this federal Act, leaving one-third still vulnerable to job loss if they have to leave work temporarily for family or medical reasons.

A CITY OF IMMIGRANTS

Paralleling the diversity in Worcester's economic base has been the remarkable diversity of ethnic groups comprising Worcester's population. In this, Worcester resembles many other American cities; the imprint of immigration on urban neighborhoods and labor markets – and the role of this imprint in the gendering of work within the U.S. – is a reminder of how gender and work are shaped at multiple scales, from the global to the local. Ever since the first Irish immigrants arrived in Worcester in 1826 to dig the Blackstone Canal, the city has attracted tens of thousands of immigrants from scores of countries around the globe. By 1865, one-third of Worcester's population was foreign born, and three-quarters of these immigrants were from Ireland (Kolesar 1989: 8), with most of the remaining quarter coming from French Canada. Twenty-five years later, in 1890, 60 percent of the city's people were of foreign parentage, but the Irish comprised less than half of Worcester's foreign stock. By that time, the Irish had been joined by Swedes, Armenians, Jews, Poles, Lithuanians, English, Finns, and Norwegians, and by 1900, immigrants from the Mediterranean region (especially Italians, Greeks, Albanians, Syrians, and Lebanese) had come to Worcester in substantial numbers (Southwick 1988; Rosenzweig 1983; Kolesar 1989).

By 1910, more than two-thirds (71.3 percent) of Worcester's population was classified by the census as "foreign stock," that is either they were immigrants themselves (33.5 percent of the city's population) or had at least one immigrant parent (37.8 percent) (U.S. Census Office 1910). Since then, Worcester has remained a city of immigrants, with a sizable proportion of its population either immigrants or children of immigrants, though the geographic sources of immigration have changed (Figure 2.7).[9] Whereas in 1910 Worcester's immigrants hailed from various corners of Europe, in 1990 they came primarily from Central and South America and to a lesser extent, from southeast Asia. The 1980s in particular saw large numbers of new arrivals from Puerto Rico and other Caribbean countries, from Central and South America, and from Vietnam and Cambodia.[10]

Although the Latino community is now the largest and fastest growing immigrant group in Worcester, the city remains overwhelmingly white: in 1980, 94 percent of the city and 97 percent of the Worcester SMSA population identified themselves as white in the census. These proportions had declined somewhat by 1990, with 87 percent of the city and 93 percent of the MSA population self-identifying as white. African-

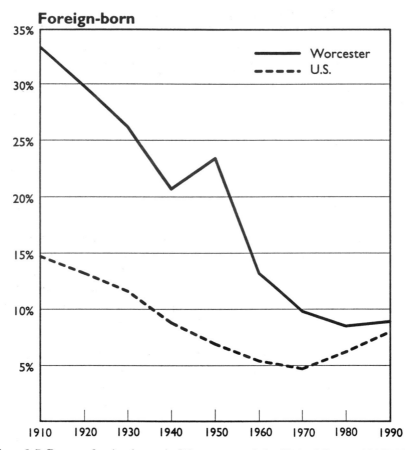

Foreign-born

Figure 2.7 Percent foreign born, in Worcester and the United States, 1910–1990
Source: U.S. Census Office 1890; 1900; 1910; U.S. Bureau of the Census 1920; 1930; 1940; 1950; 1960; 1970; 1980; 1990a

Americans have historically comprised a small percentage of Worcester's people, accounting for less than 3 percent of the city's population in 1980 and only 4.5 percent in 1990. By contrast, nearly one-tenth of the city's population was Hispanic in 1990 (Table 2.1).

Historically, attitudes toward women's paid employment varied considerably among these immigrant groups, or even, as in the case of the Irish, within them. Meagher (1986) traces these attitudes from Ireland to Worcester and attributes the enduring pattern to the abiding authority of the Catholic Church: married women were to be "sweet, good mothers," remaining at home to the exclusion not only of paid work but also of clubs and volunteer activities. Single Irish women, by contrast, were encouraged to work outside the home, and in the last two

Table 2.1 Racial composition of Worcester, 1980–90

Race or origin	1980		1990	
	MSA	City	MSA	City
White	97.0	94.4	93.4	87.3
Hispanic (any race)	2.0	4.0	4.6	9.3
Black	1.4	2.8	2.2	4.5
Asian or Pacific Islander	0.5	0.5	1.8	2.6
Native American, Eskimo and Aleut	0.2	0.3	0.2	0.4

Sources: U.S. Bureau of the Census, 1980, 1990a
Note: The problematic nature of racial categories can result in individuals classifying themselves in more than one category. Therefore columns do not add to 100 percent

decades of the nineteenth century about four-fifths of single Irish women in Worcester were in paid employment (Meagher 1986). In his memoir of Greek immigrant life in Worcester, *A Place for Us*, Nicholas Gage (1989) recounts how, in the 1950s, Nicholas's father permitted his teenaged daughter to go to work to help support the family, but only at the nearby Table Talk Pie Company, owned by a fellow Greek immigrant, who, the elder Gage knew, would keep an eye on his unmarried daughter and enforce appropriate standards of conduct. As Roy Rosenzweig notes in his careful study of the working class in turn-of-the-century Worcester, *Eight Hours for What We Will* (1983), many immigrant women worked for wages, in jobs that were segregated not only by gender but also by ethnicity: "Irish and Swedish women took the bulk of the city's domestic service positions, French-Canadian women spun and wove in the textile mills, and Jewish women ran sewing machines in the small clothing shops" (18).

A city of immigrants usually means a city of ethnic neighborhoods, and Worcester is no exception. Immigrant groups settled by nationality, often clustering around the schools, churches, taverns, and clubs they established to sustain their languages and cultures. Rosenzweig (1983: 27–9) argues that many of the immigrant groups settling in Worcester – especially the French Canadians, Swedes, and Lithuanians – pursued an anti-assimilationist strategy, at least in the early part of this century. At Assumption College, for example, founded in 1904 by Worcester's French Canadian community, French remained the language of instruction until the late 1950s. Today the remnants of the original ethnic neighborhoods linger on in street names and institutions, though the lines of ethnicity have blurred and shifted. The original Irish Catholic church on Main Street, for example, now offers services in Spanish (Plate 2.4), and Latinos now predominate on many streets where, until only 30

Plate 2.4 Originally built by Irish immigrants, St. Peter's Church now serves a large Spanish-speaking population. Photograph by John Ferrarone

years ago, French was spoken by Canadian immigrants and their descendants.

We do not mean to imply that Worcester's neighborhoods have experienced a simple process of "invasion and succession" of one immigrant group by another, as envisioned by Burgess and his Chicago School in the 1920s (likening immigrant groups to plant species). Considerable heterogeneity remains within neighborhoods, heightened by the mixture of single- and multiple-family dwellings, by the residents' varying lengths of tenure *in situ*, by the varying number of workers per household, and by the Worcester tradition of multiple (two or three) generations living under one roof in the three-apartment structures known as three-deckers (Pratt and Hanson 1988).

A "SCAB HOLE" IN MASSACHUSETTS[11]

The diversity of Worcester's ethnic quilt, together with the diversity of industries and the strength of the ethnic neighborhoods, have contributed mightily to Worcester's long tradition of antipathy toward collective bargaining in the workplace. As Roy Rosenzweig (1983: 22) observed of turn-of-the-century Worcester, "Even more remarkable than Worcester's low level of unionization was its still lower incidence of labor unrest." In 1910, only 10 percent of Worcester's workers belonged to unions, an exceptionally low figure compared to other Massachusetts cities (the 1910 unionization rate in Boston was 21 percent and in Brockton 59.7 percent, Rosenzweig 1983: 21). Although the unionization rate in Worcester in 1920 (13.7 percent) exceeded that for the U.S. in that year (12.0 percent), it is distinctly lower than the 1920 rate for Massachusetts (20.1 percent) (Figure 2.8) and is unusually low for a place with such a large manufacturing sector.[12]

Little has changed in the intervening years; Worcester remains a strongly nonunion, even anti-union, town. In 1990 Worcester's unionization rate (14.4 percent) was still below that of Massachusetts (17.6 percent) and the U.S. (16.1 percent). How did a city with such a strong manufacturing base come to embrace a staunchly nonunion culture long before the threatened departure of manufacturing jobs for other locations fanned nonunion sentiment across the country? Many have pondered this question and seen answers in Worcester's early industrial structure, ethnic heterogeneity, and corporate elite. Rosenzweig's (1983: Ch. 1) repeated assessment is that the factory owners were unified and organized, whereas the factory workers remained divided. Several ingredients – some faced by the labor movement in many places, some

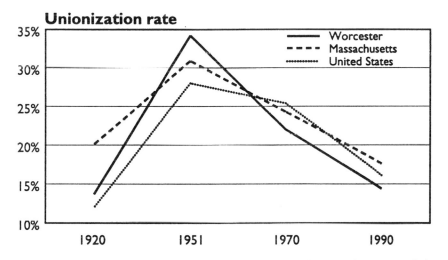

Figure 2.8 Unionization rates in Worcester, the state of Massachusetts, and the U.S., 1920–90

Source: Commonwealth of Massachusetts 1921; U.S. Bureau of the Census 1920; 1970; 1990a; Freeman and Medoff 1979; Hirsch and MacPherson 1993; Milkman 1993; Troy and Sheflin 1985

47

particular to Worcester – contributed to building this pattern. First, Worcester's diversified industrial economy meant that a strike in one industry would not succeed in disrupting the entire local economy (Cohen 1988: 165). Second, the segregation of occupations by gender meant that many of Worcester's workers had few shared interests: "male foundrymen had little in common with female corset stitchers" (Southwick 1988: 24).

Third, the work force was divided by wage and skill differentials even within occupations, so that cleavages existed within, for example, the city's machinists. Contributing to these divisions was a fourth source of variation, the diversity of immigrant nationalities. Because certain employers tended to hire workers from distinct ethnic groups, these ethnic divisions often varied by place of work. Crompton and Knowles Loomworks, for example, had a largely French Canadian and southern European work force (Cohen 1988), whereas the employees of the Norton Company (a manufacturer of ceramics and abrasives) were, for years, predominantly Swedish.

Fifth was the fact that ethnic neighborhoods and workplace divisions reinforced each other: Norton's Swedish (Protestant and Republican) labor force, for example, and Crompton and Knowles's French Canadian and southern European (Catholic and Democratic) workers lived in the neighborhoods immediately adjacent to their respective factories, which were about three-and-a-half miles apart. Workers were, therefore, not united across neighborhoods or workplaces by language, religion, or politics. Each of these five conditions was present, and hampered union organizing efforts, in other American cities besides Worcester, but the conjunction of these more prevalent conditions with several others that appear to be particular to Worcester bolstered the fortunes of anti-union forces in the city.

Rosenzweig (1983), for example, argues that the cultural cocoon of the ethnic neighborhood, particularly strong in Worcester, was the city's alternative to unions: "Yankee industrialists may have dominated the factories and the civic life of the city, but immigrants retained control over their own churches, neighborhoods, clubs, and often schools" (30). "These ethnic communities offered Worcester workers a sphere in which they could carry out a mode of life and express values, beliefs, and traditions significantly different from those prescribed by the dominant industrial elite" (27). Union organizers failed to incorporate Worcester's ethnic diversity in their strategies to enlist workers; their reliance on English to spread the union gospel, for example, neglected the linguistic divisions among their potential constituents.

In addition, some of Worcester's largest employers actively recruited skilled workers directly from Europe. Norton Company's Swedish owners, for example, drew experienced ceramicists from Skåne, Sweden

(Wahlstrom 1947); Washburn and Moen, once the world's largest maker of steel wire, recruited steel workers from Varmland, Sweden; and Whittals, an English-owned carpet manufacturer, sought skilled textile workers from Lancashire, England. These European recruits proved to be fiercely loyal to their Worcester employers and resistant to union sentiments (Cohen 1988).

The industrial, gender, ethnic, and religious diversity among Worcester workers (often reinforced by geography), therefore, hampered the fostering of union consciousness and made Worcester "far too complex a community to lend itself to the usual pattern of labor organization" (Southwick 1988: 29). Worcester's industrialists – particularly during the period of rapid industrialization, 1880–1920 – fully recognized the advantages of such divisions in the labor market (Rosenzweig 1983: 16–26) and exploited them to their own benefit. The city's manufacturers were virulently anti-union and well-organized against any attempts to unionize the Worcester work force, keeping, for example, city-wide blacklists of union organizers and ensuring that the public school curriculum included no mention of unions or economics.

Some historians (e.g., Cohen 1988; Rosenzweig 1983) see the failed strike of 1915 as the pivotal event that established Worcester's anti-union, open-shop tradition. In 1915 the International Association of Machinists, demanding pay increases and a shorter work day, launched a strike in several Worcester factories. The machine and metal trades were an important component of Worcester's economy at the time, accounting for about 40 percent of the city's industrial output in the late nineteenth and early twentieth centuries (Rosenzweig 1983: 12). The industrialists' response to the strike was swift and decisive and indicative of why unions floundered in Worcester. Union organizers were fired, and the employers refused to recognize the existence of a union. Norton's owners required all their workers to sign a pledge promising not to strike or "knowingly do anything contrary to the best interests of the Norton Grinding Company" (Cohen 1988: 163). The more-than-half the work force that refused to sign was denied entry to the factory and forbidden to communicate with the other workers. The strike ended in failure after three months (Cohen 1988: 165). Although unions did flourish briefly in Worcester at mid-century, especially among the city's steel-workers, collective bargaining was again undermined when the steel mills closed in the early 1970s; the collapse of the 1915 strike still reverberates in Worcester workplaces (Maranz 1987).

Worcester's metal industry employers used labor spies, blacklists, lockouts, and the firing of union organizers in a concerted effort to derail the city's labor movement. Facilitating this coordination among Worcester's business elite and adding to the success of their anti-union efforts was "a complex web of informal connections that . . . knit together the city's industrial elite" (Rosenzweig 1983: 14). In addition to sustaining business ties, "Worcester industrialists worshipped at the same Protestant churches, belonged to the same clubs, attended the same schools, lived in the same West Side neighborhoods, vacationed at the same resorts, and married into each other's families" (Rosenzweig 1983: 14). One of the "same schools" they attended was Worcester Polytechnic Institute (WPI), which industrial leaders had founded to ensure Worcester's leadership in the metal trades; WPI graduates, with their shared philosophy of management, filled the ranks of executives and owners of many city firms (Rosenzweig 1983: 15). These informal ties among Worcester's industrial elite not only promoted their coherent anti-unionism; they also led to local rather than national mergers taking over local firms (Cohen 1988: 170).

Sustained local ownership of the city's largest companies advanced the anti-union cause and, of course, thickened the "complex web of informal connections" that fostered anti-union solidarity among Worcester industrialists. Although some Worcester firms joined national mergers in the wave of consolidation that swept American industry at the end of the nineteenth century, for most Worcester businesses, reorganization at that time entailed a local merger. Ownership and control of most large Worcester-area firms held sway in Worcester at least into the 1960s (Rosenzweig 1983: 14), and lasted for many companies well into the 1980s. One locally owned enterprise played a particularly important role in blocking the diffusion of information about unions. The local papers, *The Worcester Telegram* and *The Evening Gazette*, were, until 1986, owned by the same family that ran the Worcester-based Wyman-Gordon machinery manufacturing company (Maranz 1987) and they ensured that management's, not labor's, outlook prevailed in print.[13]

Local ownership and the tradition of owner management were integral to the approach to labor relations that most Worcester firms adopted: industrial paternalism. Explicitly embracing a family metaphor ("we are a family here"), paternalistic employers provided their workers with housing, food (as in free milk during the working day), banking services (low-interest mortgages to promote homeownership), fuel (coal at reduced rates in winter), recreation (company picnics and ball games; parties), and gifts (turkeys at Thanksgiving and Christmas) to engender loyalty to the company. A picture of Royal Corset Company employees at lunch (and its accompanying caption), originally appearing in a company brochure, effectively conveys the message that this employer cares deeply about the physical, intellectual, and spiritual well-being of his work force (Plate 2.5). Paternalism linked Worcester firms in other ways to the community as well, for example, through employers giving preference to current employees' relatives and friends in hiring and through substantial donations to local educational, health, arts, and welfare institutions (Jonas 1992).

50

Plate 2.5 Employees of the Royal Worcester Corset Co. at lunch, *c.* 1910. Originally printed in a company brochure, this photograph was titled "White tiled dining hall" and was accompanied by a caption that read, "This flashlight photograph shows over half the employees enjoying their noonday lunch in the magnificent large dining hall. In the center will be seen a victrola which furnishes the finest music each day. At the farther end are newspaper files and a complete circulating library. Beautiful potted plants adorn the windows on every side." Photograph from the Worcester Historical Museum, reprinted with permission

These forms of investment and involvement with the local community were an explicitly geographical strategy of labor control; they grew directly out of, and at the same time served to deepen, a company's roots in Worcester. Long-term pursuit of these practices wove Worcester's large employers into the fabric of the community in ways that were still evident in the late 1980s. As Jonas (1992) has shown, a successful community-based movement to block a hostile foreign takeover of the locally owned multinational Norton Company in 1990 was a direct descendant of that company's longstanding community involvement and paternalistic approach to labor relations.

POLITICAL CULTURE

In the tradition of many northeastern and midwestern cities (Johnson 1992), the entanglement of Worcester's business elite in the life of the community encompassed City Hall as well. Cohen (1988), drawing substantially upon Zeuch (1916), describes how the industrialists' allies among Worcester's political leaders aided their cause at the time of the 1915 strike and during other union challenges early in this century. George Wright, who was mayor in 1915 and himself a wire manufacturer who had locked out his own employees, did little to end the strike and was re-elected by a landslide while the strike was still in progress (Rosenzweig 1983: 20). Serving as chairs or members of all the major city boards, Worcester's business leaders – almost all lifelong residents of Worcester – have been intimately involved in the city's development for the past century. Krefetz's (1992) description of Worcester's local government–business ties in the 1970s and 1980s echoes Cohen's (1988) description of the city in the 1910s and 1920s. In Worcester as in other cities (Johnson 1992), this business elite has pushed for the development of the downtown to the neglect of the city's neighborhoods and has been reluctant to share decision making with women and minority groups. Only in the mid-1980s did the city's non-elites succeed in pushing through reform of the city government to wrest control from a City Council comprised wholly of members elected at-large (Krefetz 1992): since 1987 the City Council has no longer been elected from the city's population as a whole; five are now elected from districts, while six members are still elected at-large.

In the years before the end of the entirely at-large city council, most city councilors, as well as most members of major city boards and commissions, came from Worcester's prosperous West Side; fewer than 10 percent came from working-class sections of the city (Krefetz 1992). As the president of a major Worcester bank recollected, "The old-style business leader was someone who owned Wyman-Gordon and in his spare time got together with his buddies and decided who would be on the City Council" (Pope 1994). While a more democratic process currently governs the selection of city councilors, now, as in the past, Worcester's industrial and commercial elite are linked together through rootedness in the Worcester community and through informal school and neighborhood ties. These intertwining, mutually reinforcing interconnections – of local ownership and control of industry, of charitable involvement in the local community, of paternalistic labor relations – have been able to flourish specifically because of people's commitment to place, people's rootedness in place. Only recently has this web of linkages begun to loosen with the loss of local ownership of the city's major industries.[14]

CONCLUSION: CONTINUITY AND CHANGE

Just as contemporary Worcester mirrors recent national trends, such as the decline of manufacturing and the rise of service employment, the increase in women's labor force participation, and the persistence of occupational segregation, so too does the Worcester of the past shine through in the Worcester of the present. Though comprising a much smaller proportion of the city's population now than at times in the past, immigrants continue to shape Worcester's history. Like immigrants coming to other North American cities in the 1980s and 1990s, Worcester's new arrivals now come from different parts of the globe than they did in the nineteenth and early twentieth centuries, but their impact on the economic, social, and political life of the city, while perhaps different in kind from that of immigrants past, remains substantial. Worcester's legendary antipathy to organized labor is enshrined in local lore and lives on in the area's relatively low union-ization rates. The tangle of relationships that enmeshes the city's business elites and civic leaders in common cause continues to be sustained and strengthened by people's enduring commitment to this place called Worcester. Certainly not unique to Worcester, the alliances between business and government that have so molded the life of the city in the past continue to shape the city today.

In many ways Worcester seems an appropriate laboratory for probing the mysteries of gender, work, and space. In 1980 the proportion of the metropolitan area's working-aged women in the paid labor force (51 percent) and the proportion of the area's labor force that was female (43 percent) were identical to the proportions in the U.S., and both these proportions rose in Worcester in tandem with the U.S. during the 1980s.[15] As we describe in the next chapter, the segregation of women and men into distinctly different occupations, with distinctly different reward structures, is as fundamental a building block of the Worcester labor market as it is of the national (and international) one. Although the "woman-employing" industries of Worcester have shifted between the 1910s and the 1980s, gender remains key to defining employment opportunities for women and men in Worcester as it does nationwide and around the globe. We turn now to describing our approach to studying gender, work, and space in the Worcester of the late 1980s.

53

3

THE WORCESTER STUDY

Investigations of gender, work, and space in contemporary American cities have been hampered by a dearth of information at sufficiently fine geographical scales. Because available data sources were inadequate to the task, we shaped three data sets, all for the Worcester metropolitan area. Our discussions of gendered labor markets in the chapters that follow draw upon these three sources of information.

THREE TAKES ON WORCESTER: AN OVERVIEW

Everything known about women's work prior to our Worcester study suggested that women's employment geographies are forged at scales much smaller than the county or the entire metropolitan area, yet employment data by gender are not readily available at a fine geographic scale. This problem has severely curtailed the analytical power of social science inquiries into the role of space and place in shaping gendered labor markets.[1] The seemingly simple question, "where – within the city – do women and men work?" could not be answered.

We therefore commissioned specially tailored runs from the Journey-to-Work Division of the U.S. Bureau of the Census to find out where women and men work. These special runs, the first of our three data sets, enabled us to make maps at the census tract scale, maps showing that, in fact, women and men work in different parts of the metropolitan area. To learn about the lives and decisions behind these maps, we – together with about a dozen students – undertook a survey of some 620 Worcester-area households, selected so as to be representative of the working-age population. Early in this field study the participants declared themselves to be part of the Worcester Expedition. The moniker stuck; the interviews from the original Worcester Expedition comprise our second source of data.

Finally, to understand the processes behind the clusters of female and male employment so evident in the maps, we targeted four small areas within the metropolitan region and studied the manufacturing and producer services firms located there. The interviews with employers and workers in these firms, comprising the second wave of the Worcester Expedition, constitute our third set of data. In this chapter we describe each of these data sets; immediately after outlining the field procedures for each wave of the Worcester Expedition, we introduce first the 1987 sample of individuals and then the 1989 sample of firms. We close with some reflections and ruminations on large-scale data gathering operations like the Worcester Expedition.

EMPLOYMENT MAPS

The decision to request special runs from the Journey-to-Work Division of the U.S. Bureau of the Census grew out of our desire to map occupational segregation at a fine spatial scale. Studies previous to our work in Worcester had documented that women's work trips are significantly shorter than men's and suggested that one possible reason might be that women's workplaces are more evenly spread across the urban landscape (Madden 1981; Hanson and Johnston 1985). Available data, however, did not permit investigating this question: some data sets provide detailed information on the gender and employment charac-teristics of individuals (thereby enabling the study of occupational segregation) but release locational information only for very gross spatial units such as the central city versus the rest of the MSA (e.g., one such data set for Worcester is the 1980 Public Use Microdata Sample, or PUMS); others provide locational data at the census tract level for place of residence but lack place-of-work information on employment broken down by gender (e.g., one such data set is that from the decadal U.S. Census of Population and Housing).

The Journey-to-Work Division of the Census was able, for a price,[2] to solve our problem. Using data from the census long form, completed by 10 percent of the population, the census bureau calculated the number of people (with particular characteristics that we specified) who lived in census tract i and worked in tract j. We requested that these tract-to-tract flows be broken down by a number of characteristics describing individuals and households, including the following: sex, occupation type (defined in terms of each occupation's gender composition),[3] industry type (e.g., manufacturing, health and human services, producer services, consumer services), mode of travel to work, household and family type (e.g., single earner, dual earner). Aggregating these tract-to-tract flows by destination yields maps of the employment landscape; aggregating the flows by origin yields maps of the residential landscape. Most important for our purposes, we were able to create employment maps showing the spatial distribution by census tract of jobs in female-dominated occupations or male-dominated occupations. Some of these maps are included in the next chapter.

THE WORCESTER EXPEDITION: WAVE 1

At the same time that we were negotiating with the census bureau to design the special runs from the 1980 Journey-to-Work Files, we were laying the groundwork for a large-scale household interview survey in Worcester. We knew we wanted to get behind the maps by learning about the everyday lives of women and men, only fragments of which would be captured in the maps of workplace and residential locations. By interviewing people in their homes, we hoped to hear about how they decided where to live and where to work, what they looked for in a job as well as in a residential setting, how they managed family and work responsibilities, how they had found their current jobs and housing, and what their experiences of work had been.

In the field

The questionnaire was semistructured, made up of both open- and closed-ended questions. The questions were grouped into five sections, probing the areas of (1) employment, (2) household arrangements (housework and child care), (3) residential location, (4) marital history, including employment information on the respondent's partner/spouse, (5) background on the respondent and the household (e.g., education, parents' occupation and education, access to an automobile, income). As part of the employment section, we collected an employment history for each respondent, asking people to tell us about each of the jobs they had held, or the breaks they had taken from the labor force, going back ten years. Similarly we recorded information on residential histories over the past ten years. In both the residential and employment histories the locations of all jobs and residences – past and present – were identified by the nearest street intersection, and these points were subsequently digitized.

The goal was to collect information on a sample that was representative of the working-age population of the Worcester metropolitan area. Details of the sampling strategy are described in the Appendix. Lasting an average of between an hour and an hour and a quarter, the interviews were conducted by about a dozen research assistants, who were graduate and undergraduate students. Although the bulk of the interviewing was carried out during the summer of 1987, interviews continued through the fall of 1987 and into the early part of 1988.

Interviewers first approached the households with a letter from us outlining the purpose of the study and indicating that an interviewer would later contact the household to set up a time for an interview. Only one-quarter of the 2,048 households we initially contacted refused to participate (30.6 percent could not be found at home after three visits, 14 percent were not eligible to be interviewed because they were not part of the target population, and 30.3 percent were interviewed (see the Appendix for details)).[4] In most households we interviewed either one woman or one man, although in 85 households two adults (most often one woman and one man) were interviewed. The goals of the study dictated that the sample include more women than men (so that we would have sufficient numbers of women to be able to compare, for example, women in different types of occupations), and hence the final sample consisted of 492 women and 206 men.[5] Of these 698 people, 526 (336 women and 190 men) were employed at the time of the interview.[6]

These interviews provide a wealth of both qualitative and quantitative information about the labor market experiences – broadly conceived – of the Worcester-area population. We must stress that because we spoke with a representative sample of households, the extremes so evident in American society at large are also evident here. The random sampling strategy yielded interviews with people living in the inner city and people living in rural exurbia, people living on Salisbury Street (one of Worcester's most prestigious addresses) and people living in public housing projects. We interviewed the owner of a small pastry manufacturing company and a few weeks later, one of his workers. We talked with a woman whose husband was the owner and chief executive officer of one of Worcester's largest firms and a woman whose husband was in jail for smuggling drugs. We talked with people who were in the paid labor force and those who were not, including people who had chosen not to work (especially housewives and early retirees) and people who had lost their jobs.

Clearly a major advantage of the sampling strategy we adopted is the ability to make generalizations about the target population as a whole and also about relatively large groups within that population. A major drawback of our approach is that we are unable to say very much about particular groups of people who make up relatively small proportions of the total population. For example, because racial and ethnic minorities comprise such a small proportion of Worcester's population, our random sampling strategy yielded a sample that was only 10 percent non-Anglo, non-white.

These interviews reveal the many ways in which women's and men's experiences in the labor force are different and the role of geography in shaping and sustaining these differences. The interviews also bear witness to significant divisions among women; the two of these that draw our attention in particular are those related to occupational segregation and to class. Before moving on to describe the 1989 survey, we introduce the 1987 sample by pausing to review in some detail the dimensions of these gender divisions that mark the Worcester labor market.

Gender divisions

Despite the diversity of life circumstances represented in the 1987 sample, the weight of occupational segregation by gender smooths and bifurcates much of this variation, pressing most employed women into female-dominated lines of work and most men into male-dominated jobs (Table 3.1). The gender typing of occupations that so focused the attention of researchers concerned about "motor-minded" girls in Worcester earlier in the century remains firmly intact. Moreover, there is no sign that occupational segregation is losing its grip among younger women. In fact, a higher percentage of women in the under-40 age cohort (58.8 percent) than in the 40 and over cohort (48.4 percent) work in gender-typical (in this case, female-dominated) occupations, although the contrast between age groups is not statistically significant. In keeping with the definitions used in other studies (e.g., Jacobs 1989) and with the definitions used in creating the employment maps for Worcester, we defined a female-dominated occupation as one[7] in which at least 70 percent of the incumbents nationwide in 1980 were women. Similarly in a male-dominated occupation at least 70 percent of the workers are men. The remaining occupations are referred to as gender-integrated.

In Worcester, as elsewhere, stark gender divisions are evident not only among occupations but also among industrial sectors (Table 3.2). More than three-quarters (77.4 percent) of the workers in health, education and welfare, for example, are women, as are nearly four-fifths (78.9 percent) of the workers in consumer services. By contrast, women comprise less than one-fifth of the workforce in construction. The

Table 3.1 Gender composition of occupation types

Occupation type	Women		Men		Total	
	n	%	n	%	n	%
Female-dominated	181	(53.9)	12	(6.3)	193	(36.7)
Gender-integrated	126	(37.5)	55	(28.9)	181	(34.4)
Male-dominated	29	(8.6)	123	(64.7)	152	(28.9)
Total	336		190		526	

Source: Hanson and Pratt, 1990
Note: Female-dominated occupations are those three-digit 1980 census occupation codes in which at least 70 percent of the incumbents, nationally, were women. Male-dominated occupations are those in which at least 70 percent of all incumbents were men. All other occupations are defined as gender-integrated

Table 3.2 Gender differences in industries of employment

Industry	Percentage of workers		Percentage of gender labor force	
	Women	Men	Women	Men
Distributive services (e.g., transportation, communication, public utilities, wholesale)	41.0	59.0	4.8	12.0
Health, education, welfare (e.g., medical services, educational institutions, public administration)	77.4	22.6	40.8	20.8
Consumer services (e.g., retail stores, personal services)	78.9	21.1	22.3	10.4
Producer services (e.g., banking, real estate, consulting services)	67.1	32.9	17.0	14.6
Construction	19.0	81.0	1.2	8.9
Manufacturing (e.g., textile products, metal industries)	42.3	57.7	14.0	33.3

Source: Hanson and Pratt, 1990

right-hand panel of Table 3.2, showing the proportions of the female and male labor forces working in each industry, reinforces these patterns: about two-fifths of the female, but only one-fifth of the male, workforce works in health, education, and welfare, whereas only 14 percent of the female, but one-third of the male workforce are in manufacturing. The gender differences in both occupation and industry of employment are striking.

Although the majority of the female workforce work in female-dominated occupations, about 46 percent of employed women do not. In the chapters that follow we explore not only the gendered division that defines sex-based occupational segregation, but also the differences among women that lead some into gender-typical occupations and others into gender-atypical lines of work. As another dimension of difference, we also examine class variations among women, men, and various neighborhoods within the Worcester area.

To open up the concept of occupational segregation, Tables 3.3 and Table 3.4 provide examples of specific occupations held by women and men in each social class and show the extent to which workers in each of the gender-based occupation types are clustered in certain social classes.

Table 3.3 Social class by occupation type for women

	Female-dominated	Gender-integrated	Male-dominated	Totals
Nonskilled manual	6 (3.3) household cleaner; kitchen worker; textile sewing machine operator	19 (15.2) hand packer; produce inspector; cook	5 (17.2) janitor; parking lot attendant; taxi driver; truck driver	30 (9.0)
Nonskilled nonmanual	87 (48.1) cashier; typist; receptionist; clerk; telephone operator; waitress	27 (21.6) production coordinator; order clerk	0 (0.0)	114 (34.0)
Skilled manual	8 (4.4) lab technician; dental hygienist; radiology technician	5 (4.0) baker; health technologist	7 (24.1) supervisor-production occupations; paper hanger; telephone installer	20 (6.0)
Skilled nonmanual	61 (33.7) registered nurse; bookkeeper; secretary	46 (36.8) management related occupations; therapist n.e.c.; painter; real estate sales; supervisor – general office	5 (17.2) sales rep. wholesale; scheduling supervisor; distribution clerks; insurance sales	112 (33.4)
Professional-managerial	19 (10.5) elementary school teacher; librarian; health teacher	28 (22.4) accountant; education administrator; secondary school teacher; social worker	12 (41.4) manager-marketing, advertising and public relations; manager/administrator n.e.c.; management analyst	59 (17.6)
				335 (100.0)

Source: Hanson and Pratt, 1991
Note: Number and percentage (in parentheses) of women in each occupation type that fall into each social class, together with examples of occupations in each category. Example occupations are among those held by women in our sample

Table 3.4 Social class by occupation type for men

	Female-dominated	Gender-integrated	Male-dominated	Totals
Nonskilled manual	0 (0.0)	9 (16.4) production inspector; bus driver; hand cutter; machine operator n.e.c.	27 (22.0) janitor; truck driver; freight handler; roofer; printing machine operator	36 (18.9)
Nonskilled nonmanual	7 (58.3) hairdresser; nursing aide or orderly; health aide	4 (7.3) postal clerk; home furnishing sales clerk	2 (1.6) car sales worker; mail carrier	13 (6.8)
Skilled manual	2 (16.7) lab technician	2 (3.6) typesetter; precision apparel and fabric worker	41 (33.3) electronic technician; telephone line installer; construction supervisor; carpenter; tool and die maker	45 (23.7)
Skilled nonmanual	0 (0.0)	27 (49.1) property/real estate manager; personnel manager; small business owner; computer programmer; salesperson business services	20 (16.3) airplane pilot; supervisor of sales workers; insurance salesperson; wholesale sales rep; computer systems analyst	47 (24.7)
Professional-managerial	3 (25.0) elementary school teacher	13 (23.6) financial manager; education administrator; secondary school teacher; accountant	33 (26.8) manager-marketing, advertising and public relations; manager-/administrator n.e.c; electrical engineer; chemist; lawyer; physician	49 (25.8)
				190 (100.0)

Source: Personal interviews, Worcester MSA, 1987
Note: Number and percentage (in parentheses) of men in each occupation type that fall into each social class, together with examples of occupations in each category.
Example occupations are among those held by men in our sample

For example, almost half (48.1 percent) of the women in female-dominated occupations work in occupations classified as nonskilled nonmanual whereas not a single woman in our sample held a nonskilled nonmanual job in a male-dominated occupation. The contrast with men is instructive. Far more women (43 percent) than men (25.7 percent) have jobs that are classified as nonskilled. Far more men (42.6 percent) than women (15 percent) work at jobs classified as manual. The sheer dearth of men in female-dominated occupations is immediately apparent.[8] Whereas the bulk of women in male-dominated occupations hold professional managerial jobs, the largest proportion of men in such occupations are skilled manual workers.

These gender divisions are tellingly reflected in labor market outcomes, most notably in the wage structure[9] but also in the availability of job benefits and in opportunities for promotion. In the U.S. context, jobs come with few guaranteed benefits. All workers in Massachusetts are eligible for unemployment insurance (which pays a portion of workers' wages when they lose a job "through no fault of their own") and for workers' compensation (which provides financial benefits to employees who suffer work-related injuries). Other benefits, however, such as health or dental insurance, paid vacations, paid maternity leave, life insurance, or a retirement plan, may or may not be provided by an employer.

Women earn significantly less than men, and women in female-dominated and gender-integrated occupations earn significantly less than do women in male-dominated occupations (see Table 3.5).[10] Job benefits as well as wages vary along gender lines and, among women, by the gender composition of their occupations. The availability of benefits perfectly mirrors the wage pattern: men are far more likely than women to have jobs where the employer contributes to the key benefits of health insurance and a retirement program, and women in female-dominated occupations are the least likely of all employed women to have jobs that include these benefits (only one-third of women in female-dominated, compared to 46 percent of other women did ($p = .03$); see Table 3.5).[11] When all kinds of benefits are considered, fewer than one in ten men (9.5 percent) had jobs that offered no benefits at all, whereas almost one-quarter of the women (23 percent) had such jobs.

Female-dominated work is frequently perceived as being low status and dead end as well as low wage. For the workers in the Worcester survey, these perceptions are well grounded in the realities of the workplace. As a measure of occupational status we used the occupational prestige score developed by Stevens and Cho (1985); this measure, the total socioeconomic index (TSEI), is based on the income and educational attributes of the total (male and female) labor force.[12] Perhaps not surprisingly, the average TSEI of men's jobs is significantly higher

Table 3.5 Labor market outcomes

| | Women | Men | Women in occupations | | |
			Female-dominated	Gender-integrated	Male-dominated
Average hourly wage	$9.34	$15.45	$8.70	$9.79	$11.70
(SD)	($5.85)	($8.80)	($5.63)	($6.07)	($5.95)
Median hourly wage	$8.60	$13.30	$7.76	$8.70	$10.30
Percentage with health insurance benefits from own job	52.1	83.2	47.3	57.3	60.3
Percentage with retirement benefits from own job	46.9	63.7	42.5	53.2	46.7
Percentage with both health insurance and retirement benefits from own job	39.0	63.7	33.5	46.0	43.3
Average occupational prestige score	39.2	43.3	33.8	44.7	48.1
(SD)	(18.8)	(21.0)	(15.7)	(19.5)	(23.1)
Percentage seeing possibility of promotion	43.4	55.1	36.8	49.6	56.7

Source: Personal interviews, Worcester MSA, 1987

($p < .05$) than that of women's jobs, and among women, gender-typical occupations have the lowest TSEI of all ($p = .000$) (Table 3.5).

To understand people's perceived chances of on-the-job advancement, we asked, "Is there another job or job title at the place you work that you would see as a promotion?" The majority of women (but less than half of the men, $p = .01$) said they could not (Table 3.5). Moreover, the proportion of women who could identify a job or a job title that would be a step up varied significantly by occupation type, with women in female-dominated occupations being the least able to do so ($p = .03$; see Table 3.5).[13] The same pattern is repeated in every row of Table 3.5: the stamp of occupational segregation brands the gender divide as well as divisions among women; it is felt in job remuneration and job prospects.

Of the employed women we interviewed, eight out of ten were married or living with a partner (77 percent of the men were), and more than two-thirds (69 percent) had at least one child living at home. Neither marital status nor presence of children in the home is, however, related to the gender composition of a woman's (or a man's) occupation or industrial sector. By contrast, both having *preschool* children at home and working part time as opposed to full time (two factors that are themselves closely linked)[14] do significantly increase the probability that a woman will work in a gender-typical occupation. Almost two-thirds (63 percent) of the women who have a child under age 6 at home work in female-dominated occupations, compared to "only" half of the women without small children at home ($p = .08$). Slightly more than one-third (36 percent) of the employed women in the sample worked part time, defined as working thirty hours or fewer a week, and working part time is strongly associated with working in a gender-typical occupation: two-thirds of part-time workers are in female-dominated occupations compared to only 46 percent of the women who work full time ($p < .01$). We explore these relationships more fully in Chapters 5 and 7.

As we noted in Chapter 1, debates have raged over the causes of gender-based occupational segregation. Because some theorists have argued that the "human capital" that individuals bring to the labor market in the form of education and work experience lies at the core of occupational segregation, we conclude this introduction to the Worcester sample with a brief overview of these key human capital variables. Although the average number of years of formal education is about the same for women (13.7 years) as it is for men (14.1), a gendered breakdown by different levels of education does reveal significant gender differences. In particular, higher proportions of working women have completed more high school and some college, whereas men were more likely than women to have done some post-graduate work (Table 3.6). Similarly, women in the three occupation groups differ in their levels of formal education, with women in male-dominated work being the least likely to have less than 12 years and the most likely to

Table 3.6 Education levels: percentage of respondents in each group by the highest level of formal education achieved

			Women in occupations		
	Women (n = 335)	Men (n = 190)	Female-dominated (n = 180)	Gender-integrated (n = 125)	Male-dominated (n = 30)
Less than 12 years	6.3	8.9	5.0	8.8	3.3
High school graduate	41.8	33.2	48.3	33.6	36.7
Some college	22.7	18.9	22.8	23.2	20.0
College graduate	18.2	19.5	18.3	19.2	13.3
Post-graduate work	11.0	19.5	5.6	15.2	26.7

Source: Personal interviews, Worcester MSA, 1987
Note: Contrast between women and men: $X^2 = 10.5$, $d.f. = 4$, $p = .03$
　　　Contrast among three groups of women: $X^2 = 20.3$, $d.f. = 4$, $p = .01$

have more than 16 years of formal schooling. It is interesting to note that the differences among women are starker than those between women and men. We conclude as do others (e.g., Jacobs 1989; Tomaskovic-Devey 1993) that the magnitude and nature of gender differences in education are not sufficient to account for the deep gendered divisions that pervade the labor market in Worcester and elsewhere.

Men in the Worcester sample also have more work experience than do women: on average, men have more years in the labor force (11.6 years) than women (9.0 years; $p < .01$)[15] and have been with their current employer about three years longer than women (nine years for men versus six years for women; $p < .01$). These differences cannot, however, explain occupational segregation because women in female-dominated occupations did not bring significantly less labor market experience to their jobs than did other employed women. More than half (55 percent) of the currently employed women had taken a break from paid work at some time during the previous ten years, but the difference in the proportions of women in gender-typical work (57 percent) and other women (51 percent) is negligible. Moreover, among those women who had spent time out of the labor force in the previous ten years, women in female-dominated occupations do not exceed the other women in length of time out.[16]

The gender divisions, as well as the differences among women, that we have described here also have a spatial expression, which we describe in the next chapter. To explore employers' roles in shaping the gendered geography of employment, we undertook the second wave of the Worcester Expedition.

THE WORCESTER EXPEDITION: WAVE 2

In the summer of 1989, and again with the help of many students, we interviewed employers and employees located in four small areas within the Worcester region. These were areas where a high proportion of the employment was in manufacturing and producer services, types of industries that are likely to be sensitive to the characteristics of the local labor force in making their location decisions.[17] Two of the areas were suburban locations and two were within the city of Worcester. Our goal was to interview forty firms in each area.[18]

The areas differed substantially from each other in terms of the class and ethnicity of their resident populations and in terms of their industrial histories. One purpose of the employer interviews was to learn about firms' locational strategies and particularly the extent to which the characteristics of a local labor force drew employers in these industries to particular parts of the Worcester area. We were also interested in learning about how the production process within the

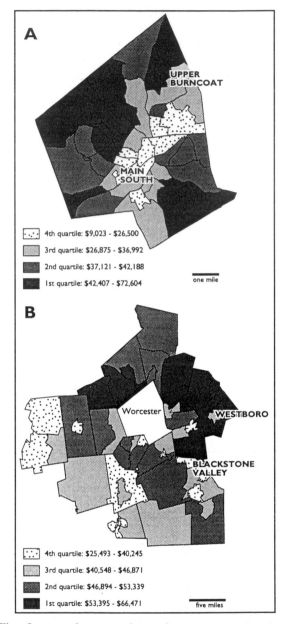

Figure 3.1 The four study areas, located on a map showing median 1990 household income by census tract. (A) Main South and Upper Burncoat are the two areas within the city; (B) the Blackstone Valley and Westborough are located in the suburban ring.

firm was shaped by the nature of the locally available labor force and how employers went about searching for and hiring workers. In view of the fact that the unemployment rate in Massachusetts and the Worcester area was below 4 percent in the summer of 1989, employers frequently reported difficulties in meeting their labor needs.

As part of the 1989 Expedition we also interviewed some employees in a small number of firms within each of the four areas, obtaining about fifty employee interviews per area. The goal of these interviews was to explore the linkages among workers in a firm as well as the ties between people's residential neighborhoods and their workplaces. The conditions of work, opportunities for advancement, the workers' residential and work histories, and their job information networks were also explored.

We provide a sketch of each of the four areas, beginning with the city and moving to the peripheral areas (Figure 3.1). By design, the four areas are distinctive. Within the city, one area, Main South, is in the old industrial core whereas the other, Upper Burncoat, lies against the northeastern city limits. Of the two suburban areas, one, the Blackstone Valley, has a centuries-long industrial history as the site of textile mills. The other, Westborough, just a couple of decades ago a small town, is

Plate 3.1 The Royal Worcester Corset factory, converted into luxury apartments in the late 1980s, is now The Royal Worcester Apartments (compare with Plate 2.1). Photograph by John Ferrarone

now a fast-growing locus of "Edge City" office buildings and industrial parks that have attracted numerous high-technology firms. In the brief depictions that follow, we seek not only to draw distinctions among the four areas but also to recognize the diversity within each as well.

MAIN SOUTH

Adjacent to the downtown, Main South is an inner-city neighborhood built up in the late-nineteenth and early-twentieth centuries. Huge red brick factories that once teemed with the industrial life of the city still line the railroad tracks, factories that manufactured carpets, textile looms, shoes, and Worcester's famous corsets. Some of this industrial space is now abandoned and boarded up, some, like the corset factory (once the home of the Royal Worcester Corset Company) has been converted to luxury apartments (now the Royal Worcester Apartments) (Plate 3.1), and some, now the cheapest industrial space in the metropolitan area, has been subdivided to house numerous small-scale and start-up manufacturing operations (Plate 3.2).

Plate 3.2 Factory buildings in Main South that were once the Crompton and Knowles loomworks are now the site of numerous smaller enterprises.
Photograph by John Ferrarone

68

Table 3.7 Characteristics of firms in the four study areas

	Main South (n = 43)	Upper Burncoat (n = 15)	Blackstone Valley (n = 30)	Westborough (n = 42)	p-value
Firm age (in years)	49.6 (47.5)	52.2 (41.5)	32.4 (36.8)	35.8 (31.1)	.15
Number of years at this location	22.7 (31.3)	21.0 (20.3)	23.2 (31.5)	9.9 (12.8)	.07
Number of locations of this firm and its branches	2.4 (5.3)	7.3 (11.8)	4.0 (8.8)	14.1 (26.2)	.01
Number of employees (this site)	59.4 (119.3)	183.8 (290.3)	57.8 (55.3)	72.4 (100.8)	.01
Number of years in Worcester area	38.1 (17.7)	32.3 (14.4)	35.2 (14.8)	24.5 (18.6)	.05
Number of previous locations in the Worcester area	1.0 (1.0)	1.0 (1.0)	0.5 (0.7)	0.6 (0.8)	.09

Source: Personal interviews with Worcester-area employers, 1989
Note: Each figure given is the average for interviewed firms in the area; numbers in parentheses are standard deviations. The *p*-value is from an analysis of variance, comparing the means of the four areas

Plate 3.3 Built as working-class housing within walking distance to the factories, three-deckers line Lovell Street in Main South. Photograph by John Ferrarone

69

The three-deckers that were built to house workers within walking distance of their employers now provide some of the lowest-rent housing in the city (Plate 3.3). Farther away from the factories but still within Main South are spacious single-family homes built at the turn of the century by Worcester's prosperous upper-middle class and now mostly converted to low-rent apartments (Plate 3.4). Once home to the Irish and later to French Canadians, Main South has become the point of reception for a large number of Spanish-speaking immigrants mostly from Puerto Rico.

This is Worcester's lowest-income neighborhood and the population is predominantly working-class, but the smooth shading in Figure 3.1

Plate 3.4 Single family houses built for members of Worcester's upper middle-class at the turn of the century are now multi-family dwellings in Main South.
Photograph by John Ferrarone

Table 3.8 Percentage of each type of firm in study area

| | *Type of firm* | | | | |
	Local (Worcester)	*Regional (New England)*	*National*	*Multi-national*	*n*
Main South	76.7	16.3	7.0	0	43
Upper Burncoat	43.8	12.5	31.3	12.5	16
Blackstone Valley	58.1	16.1	16.1	9.7	31
Westborough	44.2	9.3	18.6	27.9	43

Source: Personal interviews with Worcester-area employers, 1989
Note: $\chi^2 = 23.7$, *d.f.* = 9, *p* < .01

masks considerable diversity. Many of the long-term residents, now "empty nesters," are of French Canadian ancestry, and immigrants from Southeast Asia as well as from the Caribbean live here. Main South is also the home of Clark University, whose 2,500 students contribute yet further dimensions of diversity to the neighborhood. It is a part of the city that, prior to district representation on City Council, in 1988, had gone for a quarter of a century without a direct link to City Hall.

The firms that are located here are generally older than the firms we interviewed in Westborough and the Blackstone Valley, and they tend to be locally oriented: firms in Main South have distinctly fewer branch locations than do firms in the other three areas (Table 3.7), and multi-establishment enterprises located here are far more likely to be local in the sense that all their establishments are within the Worcester area. Not a single Main South firm in our sample is a multinational (Table 3.8). Among the Main South firms we interviewed are machine shops and metal fabricating companies, a mailing service, a commercial laundry serving mostly hospitals, thread companies, and several suppliers of temporary industrial workers.

Upper Burncoat

While sharing a city address with Main South, Upper Burncoat has an ambience that is more suburban than inner-city. Located at the northeast periphery of the city, Upper Burncoat is an area of tidy, modest single-family homes, housing the neighborhood's largely white middle-class residents (Plate 3.5). It is also the location of Great Brook Valley, the city's largest and most highly stigmatized housing project,[19] which is

Plate 3.5 A middle-class neighborhood in Upper Burncoat. Photograph by John Ferrarone

sealed off from the rest of the neighborhood by a series of roads, open space, and industrial areas (Plate 3.6 and Figure 3.2).

An example of the durable marking of space for the marginalized and the impoverished, Great Brook Valley occupies the site of the city's old poor farm, established at that location around 1850 (Nutt 1919) – a location that has remained peripheral to and isolated from the rest of the city. Known as the Home Farm, this place was where the city operated a piggery that was, until 1932, the city's only means of garbage disposal (Erskine 1981). Since the early 1950s federally sponsored low-income housing has been built in this area, which is now a public housing complex of some 1,000 apartments. The present population is largely Latino (52 percent in 1990) and includes the city's highest concentration of female-headed households (47 percent of the households living there in 1990).

The industrial park adjacent to Great Brook Valley was established in 1965 by the Worcester Business Development Corporation (WBDC) on city-owned land. An example of the alliances forged among the city's business, industrial, political, and civic leaders, the WBDC acquires land to sell or lease to industrial or commercial enterprises. The WBDC was created in the mid-1960s to ensure that businesses displaced from down-

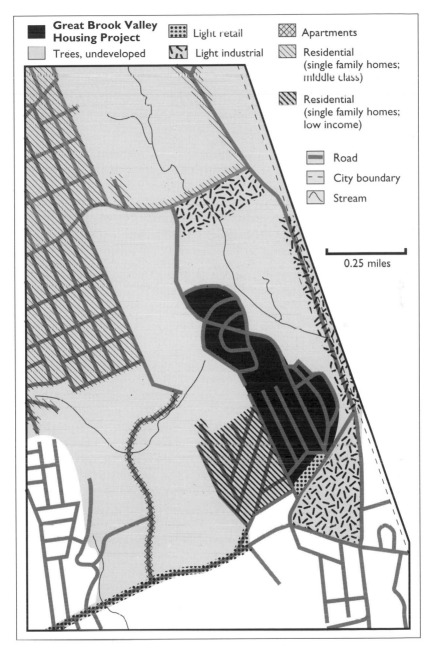

Figure 3.2 Comprised of some 1,000 units, the Great Brook Valley public housing project is sealed off from the surrounding neighborhood by open space, roads, retail shopping facilities, and industrial spaces

town Worcester through urban renewal would remain in the city. The industrial park in Upper Burncoat was one of the first to be assembled, and an early tenant was Sprague Electric, a large microelectronics firm. Not a refugee from downtown Worcester, but a firm headquartered in western Massachusetts, Sprague was looking for a largely female and unskilled but trainable workforce: when Sprague announced its decision to locate in the Upper Burncoat industrial park, a company executive was quoted in the local newspaper as saying:

> About 60 percent of Sprague's employees will be women. . . . They must have good eyesight and a high degree of manual dexterity, but the company will train them. The remaining 40 percent of workers will be men, many of them electronics technicians who must have the equivalent of a degree in electronics or chemistry from the Worcester Industrial Technical Institute at Boys Trade High School.
>
> (Worcester Telegram 1965)

Several of the Upper Burncoat firms we interviewed, including Sprague, were located in the industrial park. The number of manufacturing and producer services establishments in the Upper Burncoat area is not large (we managed to interview only sixteen),

Plate 3.6 Housing units in Great Brook Valley. Photograph by John Ferrarone

Plate 3.7 This large insurance company, Allmerica Financial, moved from downtown Worcester to its Upper Burncoat location in 1956. About 3,000 people are currently employed at this site. Photograph by John Ferrarone

although the presence of several enterprises with very large labor forces in this area means that employment in these industries is substantial here (Plate 3.7). Among the Upper Burncoat firms we interviewed were a large insurance company, and manufacturers of valves, computer components and electronics, chains, plastics, and women's clothing. Like Main South, firms in Upper Burncoat are older than those in the Blackstone Valley and Westborough (Table 3.7); but far fewer of the firms here than in Main South are local companies (less than half were); one-third were national and two were multinationals (Table 3.8).

The Blackstone Valley

Although comprised of a string of towns[20] located at sites of water power along the Blackstone River, the Blackstone Valley is an area with a highly visible identity within the Worcester region. Often simply referred to as "the valley,"[21] the Blackstone Valley is the site of the oldest mechanized production in the United States. Ever since Slater opened his spinning mill here in 1790, the economy and social life of the Blackstone Valley has been shaped by the textile industry.

In addition to cotton and wool mills, the portion of the Blackstone Valley just south of Worcester was long dominated by two large manufacturers of textile machinery, Whitin Machine Works (Plate 3.8)

Plate 3.8 Built in the mid-nineteenth century, the Whitin Machine Works straddles the Blackstone River; in the 1940s about 5,500 people were employed here building textile machinery. Photograph by John Ferrarone

and the Draper Company. Blackstone Valley workers were largely non-unionized well into the 1930s, and their wages were roughly one-half to three-quarters those of other textile workers in New England (Reynolds 1990: 179). World War II fed the last wave of prosperity in the Blackstone, but the exodus of textile firms from the Valley to the southern U.S. that had begun in the 1920s continued once the war-time boom collapsed. Once seen as the cradle of American industrial-ization, the Blackstone Valley has recently been called "the cradle of deindustrialization" (Reynolds 1990: 177).

The community-based paternalism of the mill owners is legendary. Many Blackstone Valley mill communities were company towns, where the mill owners provided housing (Plate 3.9), built schools, town halls, libraries, and fire stations, promoted athletic and cultural events, "painted homes, mowed lawns, and even changed light bulbs" (Reynolds 1989: 5). As Blackstone Valley historian Douglas Reynolds puts it, "The mill owners' commitment to their communities was neither false nor conspiratorial; instead it was pragmatic and expedient," aimed at ensuring a stable and loyal workforce. Mill owners were also successful at preventing other

Plate 3.9 A street in the Blackstone Valley town of Whitinsville lined with some of the 1,000 houses originally built during the 1880s by the Whitin Machine works to attract Armenian, Dutch, and Polish works. Photograph by John Ferrarone

firms from moving into the area: "For years the Blackstone Valley [was] a collection of mill towns, governed by strong industrial families. Other employers weren't recruited because the families who dominated the region didn't want the competition" (Newton 1985).

An enduring problem for the Blackstone Valley economy has been the Valley's poor accessibility, engineered at least in part (according to local lore) by the area's industrialists. While the Massachusetts Turnpike cuts across the top of the Blackstone Valley just south of Worcester (Rte. 90 in Figure 2.1, p. 30), no interchange was built (although one is now planned) to connect the Turnpike to Route 146, the major route threading southeast through the valley. A newspaper headline underlines the extent to which the economic fortunes of the valley are seen to hinge on improved highway access: "Valley awaiting bloom of growth: all depends on Mass Pike connector" (Godfrey and Kievra 1990).

Mindful that the poor state of transportation in the Valley helped to sustain their control over their localized workforces, local companies were supposedly instrumental in preventing the improvement of

77

east–west roads, thereby ensuring that each employer could retain a monopoly over his own local labor force (Purcell 1989; Newton 1985). As a director of the Blackstone Valley Chamber of Commerce has put it, "fifty years later and there still is no east-to-west road" (Newton 1985). In addition, Route 146 was, until significant improvements were made in the late 1980s, a narrow, winding, poorly maintained two-lane highway that was locally known as "death road." Employers succeeded, therefore, not only in keeping out other companies that might compete for their workers; they ensured that poor transportation within the Valley would make it difficult for local labor to commute elsewhere. A selectman in a Blackstone Valley town bemoans the continuing poor accessibility in the valley: "Everything comes around the valley, along Routes 495 and 146 but we seem to be left out. Everything is kind of passing us by" (Sinacola 1991).

The resident labor force in the Blackstone Valley remains largely blue collar, working class, reflecting the area's industrial history. One of the employers we interviewed referred to the area as "the Appalachia of New England" and characterized the local workers as "lower class." In recent years the low cost of land and labor and the perception that Valley residents have a strong work ethic have combined to attract many small start-up firms to Blackstone Valley towns. As a result, despite the force of industrial history here, and as testimony to the depth of deindustrialization, the firms we interviewed were on average only a little more than 30 years old (Table 3.7). Like Main South, the bulk of the firms (58.1 percent) are local and very few (only three) are multinationals (Table 3.8). Among the firms we interviewed here were textile mills, a commercial cleaning service, and manufacturers of chemicals, plastics, wire and cable, and envelopes.

Westborough

Less than ten miles to the northeast as the crow flies, Westborough offers a stark contrast to the Blackstone Valley. Whereas a distinguishing characteristic of the Blackstone has been its poor connection to the local highway network, Westborough residents enjoy excellent access to eastern and central Massachusetts: Westborough is located at the junction of Route 495 (Boston's outer beltway) and the Massachusetts Turnpike and also the junction of Route 495 and Route 9 (see Figure 2.1, p. 30). Since the Route 9 and Massachusetts Turnpike interchanges with Route 495 were completed in the late 1960s, Westborough has been transformed from an archetypical New England small town to an archetypical case of "Edge City" (Garreau 1991).[22] While Blackstone Valley business and civic leaders have been preoccupied with managing decline, the attention of Westborough residents has been riveted on managing rapid growth.

Plate 3.10 The Westborough Technology Park is an industrial park located at the intersection of Routes 9 and 495; the park houses a number of firms manufacturing computer hardware and software. Photograph by John Ferrarone

Until the mid-1960s, Westborough was a small New England town with a population of fewer than 10,000 people. Located near the center of town, the dominant industry and largest employer was Bay State Abrasives, the world's second largest manufacturer of grinding wheels (second to the Norton Company in Worcester, from which Bay State Abrasives was a spin off) (Allen 1984). Other industries had flourished for a time in the downtown here, including the manufacture of straw hats, shoes, bicycles, and women's lingerie.

Since the early 1970s, office and industrial parks have sprouted on the once-agricultural land surrounding the Route 495 interchanges. These industrial parks have attracted dozens of technology-intensive industries, particularly those in computing hardware and software (Plate 3.10). Several large "back offices" have also located in Westborough, starting with the opening in 1965 of a New England Electric Company facility to handle the firm's billing and accounting functions. High-technology firms have been drawn to Westborough and elsewhere along Route 495 because rents there through the mid-1980s were 10 percent to 20 percent lower than were rents in the Boston/Cambridge area, because the Route 495 corridor taps a highly educated labor force, and because housing prices in "Metro West" (as the western portion of Route 495 is

Plate 3.11 Westborough residential neighborhoods include some that were established in the 1950s (top) and in the 1990s (bottom). Photographs by John Ferrarone

locally known) have been significantly lower than those in Boston (Diesenhouse 1991).

Westborough residents are largely white and middle class; 95 percent of the town's 14,000 residents in 1990 were "white" with almost half of the "nonwhite" population of Asian descent, principally from India and Pakistan. Since the first residential subdivisions were carved out in the 1950s, many more have followed, their constituent houses growing ever more palatial over time (Plate 3.11). The several large apartment complexes that were part of the early-1970s building boom, unleashed a citizen backlash and prompted the passing of a zoning bylaw that effectively dampened further apartment construction by requiring that all further multi-family housing developments be approved at town meeting.[23] Westborough schools enjoy the reputation of being among the best (and best-funded) within the region; in the early 1980s about three-quarters of the town's high school graduates went on to college (Allen 1984). Other town services like the library, the senior center, and the recreation program have continued to flourish in Westborough in contrast to many other central Massachusetts towns, which have been facing fiscal austerity. Superb access to high-paying jobs, a relatively new housing stock, and high-quality services combine to make Westborough among the most sought after residential environments in the Worcester area.

Among the Westborough firms in our sample are computer hardware and software companies, back offices, a company providing trade show services, and manufacturers of abrasives, hydraulics, and precision air bearings. The Westborough enterprises we interviewed are larger (in terms of the number of locations/branches) and are newer to the Worcester area and to their present locations than are establishments in the other study areas (Table 3.7). Westborough firms tend to be either local (44.2 percent) or multinational (27.9 percent) (Table 3.8). In keeping with Westborough's Route 495 location, a large segment (44 percent) of the Westborough firms in our sample made high-technology products.

The preceding account of the design of the Worcester Expedition and the accompanying introductions to the 1987 and 1989 samples have followed a well-trodden path in "conventional" social science. In the retelling, what was in fact a dense network of routes with off-shoots, meanders, dead-ends, and loops, becomes a linear superhighway to the "facts" and the "truth." Before proceeding to report the many sightings and discoveries made on the Worcester Expedition, we pause to reflect on the creation of knowledge and the Expedition's role in creating the knowledge we report in the remainder of the book.

LOOKING BACK AT THE WORCESTER EXPEDITION

When the students working on the project first began to speak of the "Worcester Expedition," we latched on to the term because it expressed our wonderment at stumbling into worlds that were previously unknown to us and, also, because we were funded by the National Geographic Society. In retrospect, the term "expedition" seems especially apt. It expresses our central relationship to scientific geographical traditions. But also, because of its highly marked status in critiques of colonialism, the term captures in an ironic way our ambivalence toward some of the baggage that we carried along on the trip.

This ambivalence leads us to tell two kinds of stories about the Worcester Expedition, neither of which conforms to traditional descriptions of methodology. First, we want to place ourselves as both willing participants and captives in the technology of science, and briefly consider the processes of abstraction through which our scientific facts were and continue to be constructed. What follows is only a sketch, but it establishes a context and urgency for the second narrative, which is an attempt to contextualize and localize our knowledge claims. Much of what follows was conceived through conversations with eight of the research assistants who worked on the Worcester Expedition; we quote extensively from these accounts of their field work experiences.[24]

Science networks

Latour (1987) characterizes technoscience as a network that generates information that can be brought from "the field," stabilized through a process of abstraction, and then combined with other information similarly produced. The result is both power and knowledge: the ability to dominate people, things and events at a distance, *and* the ability to see the world in new ways. It is possible to read our research through this frame; our interest in doing so increases as some of the scientific facts of our own creation extend through networks beyond our control.

All research involves abstraction and representation, and ethnographies, no less than survey methods, are caught in a web of "methodological horrors" (Woolgar 1988), such as the recognition that the links between what one wishes to represent and what one constructs as representations are both indeterminate and reflexively intertwined. Nevertheless, the rupture of abstraction might be more obvious when survey methods, semi-structured questionnaires, and quantitative methods are deployed, as they were in the Worcester Expedition, and our methods are susceptible to well-rehearsed critiques of quantitative methods (e.g., Ley and Samuels 1978). We suspect, for example, that there are large gaps in our information about child care because it was impossible to wedge it into well-defined categories or even to describe it in an unstructured but succinct way. Despite the fact that we took care designing a question about child care, one that allowed interviewers to record a variety of arrangements, we collected much less information on this subject than we expected. This may reflect the fact that child care is used infrequently in Worcester or, possibly, that arrangements are often so complex and variable that they are simply too difficult to describe and record on a structured questionnaire. Consider one informal description of child care arrangements in Worcester:

> From this hour to this hour a neighbor comes in. And then an aunt who lives across the street comes over. And then a neighbor's daughter gets home from school and comes over. And then she [the neighbor's daughter] gets her sister to come over when she goes to a lesson. And the arrangements are never the same from week to week and day to day.

This type of detail never found its way into our completed questionnaires; indeed respondents would be unlikely to provide it, given the length of the interview.

Structured questionnaires also impose a fragmenting grid that makes it difficult to understand the complexity of meanings of, and interconnections among, different events and strands of life. This complexity comes through in this comment from one of our research assistants, who grew up in Worcester:

> Like what would you say? . . . How would I write it down if someone told me: 'My sister's boyfriend got me the job. But they had just started dating. Now he's my brother-in-law.' Something like that. That would be so common in Worcester. Or an ex-boyfriend. How do you code that? Or a relative who is also a neighbor? Which is very different from a relative who is not.

As a final point, structured interviews inevitably import an interpretive framework that allows you to see some things but may hide other relations from view. The interviewer quoted above also wondered whether we missed the work done by older siblings:

> [it] was very difficult for that to come through, the way that the interview was structured. We asked about household chores in terms of 'is it either you or your spouse or someone else who does it?' The someone else sounded like it was outside the family. But among my friends it was very common for the oldest child to make dinner.

We troubled over this process of abstraction and tried to bridge it in a number of ways. We encouraged the interviewers, for example, to stray from the interview schedule when other themes emerged, and to jot down extended quotes when they seemed to be of interest. When we came to code the unstructured interview questions, we developed an elaborate system of categories. Responses to the question: "How have your job priorities changed?" spawned no fewer than 48 categories:

"Money more important now because partner sick" was, for example, distinguished from "Money more important now because partner died," and so on. There was, of course, a certain naiveté associated with this attempt to capture the specificity of meaning through an elaborate array of codes. These codes were a step toward quantitative analyses, and the use of statistical analyses as a means of finding relationships dictates the collapsing of categories to ensure adequate numerical representation in each. As our interpretation of the interviews proceeded, we recognized the futility of trying to retain the specificity of meaning through the proliferation of codes and were troubled by the loss of contextual information that necessarily accompanies quantitative analyses. We began to revisit the questionnaires and to treat the extended quotes written down by interviewers as texts. Given the methods employed (e.g., the interviews were not tape recorded), the written texts are limited, and it remains a continuing frustration to know that the interviewers had, in many cases, a far richer and more interesting interview experience than is evident from the written account returned to us. Hearing these silences in the written accounts, we are deeply aware that these are the *n*th order inscriptions so eloquently described by Latour (1987, 234).

Admitting these difficulties, we also want to take up another strand of Latour's argument; this is that the process of abstraction, which allows for combining diverse bits of information in novel ways, also allows you to see the world differently. There can be something very powerful about these constructed abstractions. We are struck by the power of some descriptive statistics that emerged out of our research: for example, that only one woman in our entire sample had an annual income of at least $35,000 and a school-aged child while one out of every three men in our sample earned this much. Equally striking, no man in our sample found his job through a work-related or community-based contact that was a woman. These statistics would have no impact had they not emerged from a large, randomly selected sample.

Drawing on Latour, however, we cannot help but ask, powerful for whom and in what context? Certainly we can be (and have been) criticized for our objectifying gazes, and for our position within academic "centers" of power. A research assistant who worked on the Worcester project writes:

> [Most] disconcerting was the exclusion of the survey participants from the survey design. The survey was designed to answer questions of interest in the academic literature. The unidirectional flow of information from the community to the university rarely benefits the community.[25] What about the concerns of those living in the community? Who answers their questions? University researchers

84

are well equipped to provide the community with this kind of intellectual infrastructure, but they rarely do.

At the end of a research presentation one of us has been accused of exploiting the women we interviewed and advised to "[s]top looking at [the exploitative practices] of employers and look at yourselves."

These are serious charges, and our intention is not to lay them to rest but to add some complexity to our understanding of them. It is undoubtedly true that our surprising facts are surprising to us because of our location as academics. It is also true that we are funded as academic scholars and write as, and for, academics. A research assistant tells of bringing our published academic work to his sister, a life-long Worcester resident, because he thought that she might see her life in it:

> R.A.: Part of it was the wording, but she didn't see her life portrayed there. I did.
> G.P.: How did she feel about that?
> R.A: I think she felt ignorant in a way. And she couldn't see the significance of the theoretical point that you were trying to make.

Another research assistant worried that the interview experience was a profoundly depressing one for some respondents:

> Now and [then], you would interview someone who perhaps came out of the process depressed. I remember one woman saying: "Yes, all my life," noting that she had lived in the same house for the last 25 years and had been in the same "dull" job. She commented, "Gee, I guess my life has been pretty boring." Perhaps this is constructive, but perhaps not.

But the traces that we left in the field seem far more ambiguous in their effects than a solely negative reading of exploitative academic practices would suggest. Another research assistant spoke of the emancipatory effect of doing an interview.

> R.A: It was a privilege for me to do that interview with [that woman]. Nobody had bothered to ask her about her life. Some people said they'd never had the chance to sit for an hour and reflect. [It seems to me that in some cases] we really made [people] think about their domestic situation or their employment situation. If ever what we did was unethical, then I think that you have to think of the [positive] ethics of what we did. You have to think about the ideas that we left behind.
> G.P.: What things did we cause people to think about?
> R.A.: The housework, the distribution of household chores.

G.P.: How was that, just because they had to quantify it?

R.A.: Because they [women] had to admit that they were doing the bulk of the household work and holding down a job. You know if she's sitting there doing the interview and he's in the other room watching football you know, she says "heh, I have proof here!" Positivism can be useful. I can remember someone shouting over their shoulder "Hey, you got to do more around here" after we got through that question. Or doing the employment history. They'd get to this point where they'd have good jobs and then they'd have kids or someone would get sick or their husband would get laid off or they'd have to move. You allowed people to see these patterns.

G.P.: In seeing these patterns, would they make comments?

R.A.: Maybe they wouldn't make comments, but you could tell that they were thinking about it. You could tell that they were being reflective.

G.P.: What about the dangers of doing this?

R.A. I know what you're saying, but I guess I think it better than it never happening at all which is what I saw in Worcester. You know, like, people tend not to reflect on their lives.

In this sense, we have not only abstracted and generalized patterns but offered an opportunity for them to be pieced together locally, by individual women and men. Even this interpretation is not so simple, however; were the women who made such comments seeing themselves through our middle-class, university-educated eyes? Did they anticipate our expectations that housework should be shared equally and mediate the disjuncture between our expectations and their own lives by remarks such as, "Hey, you got to do more around here!"? These reflections do not "solve" the problem of the ethics of bringing one's own agenda to study others, but they do make more complex our interpretation of the effects of the process.

The traces from the project flow in other directions as well, for example, laterally through the academy. Most research is conceived as a product, but it is also a process, including a pedagogic one. Most of our research assistants were or are graduate students in geography. Few have used only structured questionnaires and quantitative methods in their own research. Most have opted for more qualitative methods (sometimes in concert with quantitative ones), in some cases as a result of working on the Worcester Expedition and because our research provides a base for more contextual accounts (Gilbert 1993, 1994). Nevertheless, even when the substantive focus of their own research is very different (e.g., hostel life in South Africa and local political responses to AIDS in Vancouver), research assistants tell of the many and different ways that the experience of working on the Worcester project affected their own research. For example: "It was through the project that my interest in a gendered analysis developed" or "you made me interested in 'work' – a topic that I had been thoroughly turned off by the structural Marxists that surrounded me . . . [Y]our research helped me to see work more culturally, rather than narrowly or economistically as the operationalization of 'the labor process.'" Again, there are complexities and contradictions, as semi-structured interviews and quantitative analyses spawn ethnographic accounts.

One of Latour's most useful points, however, is that networks of technoscience extend beyond the walls of the academy. To argue that the production of scientific knowledge is a social process involves more than tracing the specific connections between "principal investigators," interviewers, and interviewees. We have been interested in and concerned by the ways that our constructed knowledge has traveled through networks far beyond our control and especially in how the abstracting process that we set in motion (and try to resist) becomes exaggerated as our constructed facts are circulated in more public, media reports. Many academics disparage the distortions that occur when their research receives media attention (Roberts 1984). Dorothy Smith (1990) sees these distortions as having shared characteristics, which she locates in the process of detachment – from speaker, from specific audience, and from actual events. We can document this process of distancing and detachment in media reports of our research. It is a further step in a process of abstraction and it involves stripping specificity, allowing for bits of our research to be combined with those of others to tell a unified story.

Our research has been described in about a dozen media reports, but we will take just one example to make our point. This example appeared on the front page of the *Wall Street Journal* in a regular feature called the Labor Letter (2 February 1993). The report of our research appeared under the heading, "Breaking Barriers: Work-force diversity faces big obstacles." The first paragraph of a longer report was given over to a description of our research and reads as follows:

Many employers seek labor uniformity by recruiting largely through word-of-mouth or localized advertising, say geographers Susan Hanson of Clark University and Geraldine Pratt of the University of British Columbia. In Worcester, Mass., they found companies target specific types of workers, such as Polish immigrants or working-class women, in deciding where to locate plants.

There is much of interest in this short passage. There is, for example, a confusion of cause and effect: insofar as we have argued that employers' hiring strategies often have the effect of creating labor uniformity but that the causes are more varied and complex. Agency is also lodged with the employers (while we have been explicit about a "dynamic dependency" between employers and employees). Moreover, employers' recruitment strategies are tied to specific, seemingly vulnerable, populations (immigrant Poles and working-class women) when in fact many employers tied their locational decisions to the accessibility of skilled (male) workers and professionals. Especially noteworthy is the radical loss of specificity, for starters, that we interviewed only employers in producer services and manufacturing industries and only in four areas of Worcester. Stripped of specificity, our study is easily packaged with another, which purportedly claims that companies that are trying to promote diversity are threatened by "backlash." The authors of this second study apparently advise companies to assign more white males to conduct diversity training so that employees won't dismiss diversity as being "only about women and minorities." It is alarming to find one's work as the headliner for a plea to place more white males in charge of diversity training. In this instance, abstracted facts, radically reduced and decontextualized, are combined to tell a story about the difficulty of breaking barriers because of white male employers' agency and the necessity for white males to reclaim this process.

Localizing knowledge claims

Our second story is about messiness and disorder. There are two strands to it. First we want to describe a certain wildness and unaccountability that was part of our research process. Second, we want to attempt to understand how some of our interpretations and silences result from how we as producers of knowledge are/were situated within networks of social relations.

The first strand deserves a fuller treatment than it will receive here because it opens up an issue around which there is an astonishing silence: the use of research assistants to "collect data." Many anthropologists have troubled over what has been called the "double mediation" that occurs when anthropologists construct their ethnographic accounts, namely the fact that informants create a representation of themselves for ethnographers, who then – in relation with informants – create another representation (e.g., Okely and Callaway 1992). Despite this type of discussion and despite the fact that almost all of us use research assistants at one time or another, there is almost no discussion of what we do when we hire research assistants, what assumptions lie behind it, what the effects are, and how the relationship is and might better be conceived and put into practice. (The one exception that we know of is Kay, 1990.) This is a politics and practice that lies very close to home and we ignore it.

That we ignore the extra layer of mediation introduced by research assistants may reflect an unacknowledged acceptance of a particular and pervasive "ideology of representation," to use Woolgar's (1988) terminology. This ideology assumes a type of passivity in relation to the assumed facts of the world. It assumes the idea that anyone would find the same facts if they went to look for them (using the same methodology) and a certain interchangeability of human bodies that runs directly counter to prevalent ideas about the constructed nature of and situatedness of knowledge claims.

Recognizing this, we have to acknowledge a radical inability on our part to account for our research process. Our efforts to situate our knowledge are made problematic by the fact that the mediation of representation flowed not only through our two bodies and those of interviewees, but also the bodies of more than 25 other people – the research assistants – who differed in gender, nationality, ethnicity, sexual orientation, knowledge of and experience in Worcester, and in many other ways that we do not know. In an effort to fill in the blank and to piece together a partial reconstruction of the active role played by research assistants in constructing our knowledge claims, we contacted many of them and asked them to talk about their experiences.

The conversations with our research assistants revealed them to be anything but passive and transparent conduits of facts. Research assistants were very active producers of the texts that they presented to us as completed interviews. We will give just two examples. One interviewer was reflective about the way in which she actively constructed herself in different ways in different interview situations:

> At a small Main South electronics firm, we'd obtained the interview through Clark University connections, I think. So in that case, then, we were very much positioned as "Clark students." In other situations, I attempted to construct "outsiderness" (to Worcester) as a means of "probing" the question of tight local labor markets – i.e., this difficulty of finding workers seemed very different from Vancouver so could they "tell me more" about it.

There are many issues, including ethical ones, opened up by this recognition that interviews are conversational performances. We wish to draw the simple point that it flags the fact that those being interviewed were no doubt also doing this, in ways that are impossible for us to understand. This is an issue that pursues almost every social scientist (see for example Crick's (1992) attempt to figure out the performative strategies of ethnographer (himself) and an informant), but the sheer number of performers on the stage in the Worcester Expedition makes the accountability issue particularly problematic.

As a second example, a number of interviewers told us of translation exercises that they engaged in as they attempted to offer help with the structured questionnaire. We asked those being interviewed to reconstruct their employment and residential histories, something that many people found difficult to do. Interviewers were helpful translators. One interviewer recalled offering the following cues to women: that they might try to reconstruct their histories by thinking about what they were doing when they had their various children or at critical points in their lives as mothers, for example, when their children went to school. Another interviewer recalled that: "I think that with (especially younger) men it was a case of identifying the first job they had held, rather than working backwards from the present one." What this means for the life histories collected – that some women were encouraged to structure their memories of employment through their roles as mothers, and some interviewees read their histories backwards while others read them forward through time – is difficult to say,[26] but it does signal the active and varied roles played by interviewers in the construction of life histories.

Admitting the difficulty of locating our research assistants and accounting for the myriad ways that they produced the interview texts, nevertheless we can try to situate ourselves as researchers. Despite the necessity and fashionability of situating oneself in relation to knowledge claims, it is difficult to do this, and it is important to remember that stories about one's own situation are also constructions.

Aspects of one's own situation become more evident in retrospect. Some of our blind spots simply became clearer when, as the research progressed, we began to ask questions that we could not answer with the information at hand. We became more and more interested, for example, in the tangled networks of relations in Worcester, and began to notice how these networks were used to find jobs and housing. As one of our interviewers put it: "Worcester is like a Dickens novel. Everyone comes together in the end in weird relationships." Reflecting on our research process and published accounts to date, he was critical, as a long-term resident of Worcester, of our inability to capture the fullness of these networks: "It's not only that this is how you get your job," he argued. "It's also how you get your partner. It's connected to just everything. It's the way you're treated in school. Your teacher probably taught your older brother and sister and maybe even your father and mother." Neither of us has lived the intensity of these extremely localized networks. Living within personal histories of very different time-geographies, it took us some time to recognize them and to understand their importance. (Of course, our "key informant" is also positioned within a particular neighborhood and social world within Worcester and his remarks are also constructions.)

We attempted to pursue the theme of networks more fully in the 1989 interviews, but the context of the interviews, which were carried out mostly at the place of employment, made this quite difficult. Again, this points to the constructed nature of facts. In the words of an interviewer:

> The set up for talking about networks and stuff like that would have worked better in the 1987 interviews. Usually [in 1987] there was no one else in the room. Usually it was in people's homes. You could observe them in some kind of context. Whereas in '89 they were herded in, and we were sort of paired off one to one and we'd be like in a cafeteria. Now remember, I did   a lot of the Westborough ones. In 1987 there were quite a few interruptions and sometimes the interruptions were where you got at that context. And you couldn't, there was no way at getting at that context of those networks in the '89 interviews. Whereas say you were in someone's home and the phone rang and it was this woman's sister and they were negotiating about who was going to take care of whose kids and when, you got a better sense of the networks there. . . . The employee interviews are not as rich because they were outside people's homes. They were looking over their shoulders. We were doing it in a cafeteria. There was a certain reluctance. They didn't know what they were for or how they'd be used. I remember the 1987 interviews as being far richer. You had us out in Webster or Shrewsbury months at a time. It was like a little exploration.

Our earlier blind spots, rooted very much in our own personal histories, limited the richness of the stories that we can tell about networks.

One's situation and its effect on the research process also become more apparent in retrospect because you change over the course of the research project – especially when it occurs over a long period of time, as this one has – and at the end you see things differently from the way you did at the beginning. What counts as facts is context-dependent and, as you and the rest of the world change, new events become strange and therefore assume the status of interesting "facts" (Woolgar 1988). Relatively late in our research process, we became interested in looking at how child-related breaks affect different women's job trajectories. One of the theoretical arguments that we have been concerned to make is that time out of the labor force should not be conceived as an experiential black-hole, as it is by most economists and sociologists who have looked at the issue. No doubt our sensitivity to this has been sharpened by the fact that one of us has had a maternity break in recent years. It is a point of some embarrassment, therefore, to recognize that our interviews in some ways carry the assumption that we are now keen to criticize and made it almost impossible for women who had taken extended child-related breaks to tell us about the richness of their lives. Having lived adult lives marked by quite constant labor market participation, we had no experience that allowed us to listen to women living different lives of work.

Finally, one learns to situate oneself through the help of others.[27] Others, situated somewhat differently, can help you to see how your own situation structures your assumptions and research questions, with the understanding that the need and potential for aid is reciprocal: no one stands in a position of invulnerability. In this respect, our research assistants have helped us to see aspects of ourselves that are very difficult to detect. One writes, for example:

> In some interviews, it was apparent that [your] way of thinking about decisions of where to live and work was alien. I can think of one woman who found the question very strange. The reasons why she had moved were not related to work, child care, etc., but rather, she had been thrown out of her house by her boyfriend and then in the second place she had lived, the house burnt down and so she had to move. Perhaps this reflects the class differences [between] researchers and interviewees.

Our lives had not prepared us for an interpretative framework of calamity, violence, and extreme vulnerability, and it is possible that we impose a framework of premeditated self-direction that is simply foreign to some people's lives (Laurie and Sullivan 1991 explore this point in some detail). This may result from our class positions. It may also reflect the influence of urban geography theory, which emphasizes "rational decision making" in selecting the locations of home and work place. Despite our attempts to critique this theory, to some extent we are still caught within it.

The last observation points to some directions for rethinking the relationship between research assistants and so-called principal investigators. Rather than seeing research assistants simply as a liability – the extra bodies that add another layer of distance and a large degree of unaccountability – one can also conceive of research assistants as extremely helpful collaborators.

In conclusion, we offer these observations to signal that the research we describe in the coming chapters, despite the presentation of maps, tables, and percentages, is an artful production. By insisting on the artfulness of this production, we also hope to retain some control over the use to which it is put and, in particular, to localize our knowledge claims. "Statistics . . . ," writes Anne Pugh (1990, 109) "need chaperoning." This is our attempt to escort them into the world.

4

THE FRICTION OF DISTANCE AND GENDERED GEOGRAPHIES OF EMPLOYMENT

A 1979 article in the *Worcester Telegram*, entitled "Women face struggle looking for jobs," began as follows: "Going out and looking for a job for the first time can be an awesome project – especially if you are a woman. . . . There are so many things to consider. First and foremost, of course, is what you want to do. Running a close second is what you can do – what your abilities, your skills are. And then of course it depends what's available in this geographic area" (Towne 1979). This article goes on to stress the special struggles that women, as women, face in looking for work: "your situation, if you are a woman, is not only different from a man's, it is difficult, if not impossible." Yet the problem, hinted at toward the beginning of the article ("it depends what's available in this geographic area") is never taken up. What does "available in this geographic area" really mean? The phrase suggests that women's labor force participation depends on the "availability" of "appropriate" jobs in the "right" locations. In fact, geography lies at the heart of any understanding of how women's situation in the labor market is different from (and often more difficult than) a man's. Geography is also essential to any understanding of the different labor market experiences of different groups of women.

In this chapter we begin exploring how geography shapes the gendering of work and how gender shapes the geography of employment. The key lies in unraveling that offhand yet vaguely mysterious assertion "of course, it depends what's available in this geographic area." We begin by excavating assumptions about the friction of distance that are embedded in notions of separate spheres for men ("the public") and for women ("the private") and continue by looking at how the size of "this geographic area" is different for women and men and for different groups of women. We do this by examining how long people take to travel to work. We then examine the spatial distribution of different types of jobs and consider the implications of these employment maps for gender-based occupational segregation. Throughout this chapter, the

93

emphasis is on describing patterns; the reasons for the patterns and the consequences of them are taken up in later chapters.

THE PUBLIC, THE PRIVATE, AND ACCESS TO OPPORTUNITIES

The relationships among geography, gender, and employment are implicated, highly abstractly, in the gendered ideas about "the public" and "the private" that emerged in the late nineteenth and early twentieth centuries. This dichotomy draws a distinction between suburbs, women, femininity, and "the private sphere" on the one hand and the city, men, and masculinity, and "the public sphere" on the other. "Urban life and men tend to be thought of as more aggressive, assertive, definers of important world events, intellectual, powerful, active, and sometimes dangerous. Women and suburbs share domesticity, repose, closeness to nature, lack of seriousness, mindlessness, and safety" (Saegert 1980: S96–S97). Some have seen suburbanization and the proliferation of the single-family home as part of a male plan to enforce maternal roles for women and to reinforce patriarchy by spatially separating women from "the corruptions of the world of work" (Fishman 1990: 41; see also Wilson 1991).

Feminists have roundly criticized the public/private polarity on the grounds that it overgeneralizes from a small portion of society (namely the upper-middle class) and that it portrays and perpetuates a false stereotype by ignoring the facts that large numbers of women, especially immigrants and minorities, have always lived and worked in the city and that the home has always been a site of production as well as reproduction (Sacks 1989; Nakano-Glenn 1985). Despite these criticisms and the profound recent changes that have blurred any simple urban–suburban dichotomy in American settlement patterns, an evening spent watching network TV and the accompanying advertisements admonishes us that these associations (of women with home and suburb; of men with work and city) remain firmly embedded in American popular culture.

What fascinates us about the tenacity of the public/private mindset is not only that it implies a geographic fission in life that follows gender contours (however erroneous), but also that its cedes an enormous power to space, or more specifically to spatial separation, albeit at a very high level of spatial generalization. According to this popular interpretation of suburban design, women were to be kept out of the labor force by being spatially removed to the suburbs (Wajcman 1991; Wilson 1991; Fishman 1990). Distance, then, was to be the prime instrument used in isolating women not only from jobs but also from power and involvement in the body politic.[1] Moreover, at the bottom of this argument is the idea that the male urban designers behind this plan assumed a striking gender difference in the friction of distance: although women and men would inhabit the same suburban houses, men would travel to the city to work but women would not. The same distance that would pose no problem to men would be an insuperable barrier for women and would suffice, therefore, to prevent their entry into the paid workforce.

The urban/suburban–public/private dichotomy rests, then, on the assumption of a profound difference between women and men in their ability to overcome distance and on its corollary that women and men living in the same location are not likely to have access to the same set of opportunities. "What's available in this geographic area" will therefore not be the same for women and for men, in large part because women and men – through their differences in willingness and ability to travel – will define the relevant geographic area differently.[2]

If indeed suburbanization was a male plan to encase women in a prison of distance from opportunity, how accurate was the bedrock assumption? Clearly women, and especially married women with children, have overcome the barrier of distance and entered the labor force in record numbers. Moreover, giving the lie to the notion that only "the private" resides in the suburbs, today's suburban women are more likely to be in the paid labor force than are women living in cities. On the other hand, many studies have documented women's resistance to traveling as far as men do to reach their paid employment. Julia Ericksen (1977: 429) notes that as long ago as 1907, Pratt (1911) found that women in New York City spent less time on the journey to work than did their male counterparts. We turn now to examine how far people living in Worcester in the late 1980s travel on the journey to work as an indicator of the spatial dimension of their labor markets.

HOW LONG DOES IT TAKE YOU TO GET TO WORK?

Job location matters. The recent national study, on changes in the workforce (Galinsky *et al.* 1993), which interviewed a nationally representative sample of some 3,400 workers, asked respondents how important various factors were in their decision to take a job with their current employer. "Job location" was judged to be "very important" by half of the respondents, outranking fringe benefits (43 percent), opportunities for advancement (37 percent), and salary or wage (35 percent). Yet exactly what the study's designers meant by "job location" or what the workers had in mind when responding to the "job location" prompt is not clear. The possible interpretations are almost endless: Location on the west coast? in this state? in this metropolitan area? in this neighborhood? on a mountain side? at a freeway interchange? close to home? far from home? next door to my favorite book store?

In our Worcester study we have looked at job location in a number of ways, including primarily travel time from home to work but also distance and ease of access from home and proximity to school and childcare. What is clear is that the geographic scale at which job location matters to most people and especially to women is very fine indeed. Just how fine is illustrated by the experience of a young woman from Uxbridge, a town in the Blackstone Valley, who at the time of the interview, was working as a shipper in a yarn factory five minutes (by car) away from her home. With ten years of formal education and two school-aged children, this woman had held nine different jobs (including waitress, cashier, warehouse picker, mill worker) in the previous eight years – all, except one, located within 15 minutes of home. She explained that she had always worked close to home "to be close to the kids," but the extreme proximity to home of her present job was made necessary when one of the family's two cars broke down. Her husband was now driving the one usable car to work, requiring her to ride to work with a friend. Nevertheless, she had carefully chosen her job to be within walking distance of home in case her friend "didn't show up" or did not go to work on a given day.

While job location may be "very important" to both women and men in their choice of jobs, women tend to choose jobs that are significantly

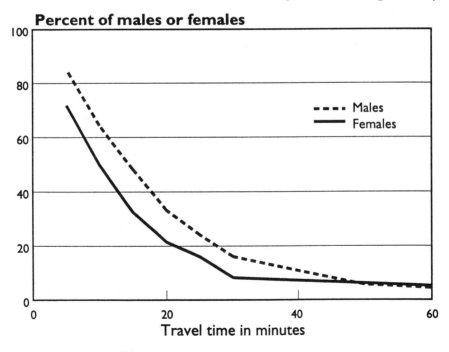

Figure 4.1 Men's work trips tend to be longer than women's
Source: Personal interviews, Worcester MSA, 1987

closer to home than are the jobs that men choose. Although one-third of the men in our sample traveled 20 minutes or more to work, only one-fifth of the women had jobs that entailed a work trip at least that long (see Figure 4.1). On average, women traveled 15.5 minutes to work whereas men averaged 20.4 minutes ($p < .01$). This overall gender difference needs to be refined because so many women work part time (35.7 percent of the women but only 3.9 percent of the men in this sample worked part time), and part-time workers tend to work closer to home than full-time workers (13.6 minutes on average for part-time workers versus 18.3 minutes for full-time workers in this sample). Because so few men (only seven) worked part time, we do not compare the travel times of part-time male and female workers, but the significant gender difference in work-trip travel times is still evident when we consider only full-time workers (see Table 4.1).

This gender difference in work-trip length in Worcester mirrors the findings from many other studies based on large random samples from other cities and other times (Madden 1981; Hanson and Johnston 1985; Howe and O"Connor 1982; Fagnani 1983). The recent work of McLafferty and Preston (1991) and Johnston-Anumonwo (1994) on the commuting patterns of racialized minorities in American cities highlights, however, how the consistency of this finding depends on the fact that large random samples drawn from North American, European, and Australian cities tend to be predominantly white. We now know that this frequently documented gender difference holds only for whites and that

Table 4.1 Journey-to-work travel times

Group	n	Travel time (minutes) Mean	SD
Women			
Full-time employed	210	16.8	14.3
in female-dominated occupations	97	13.3	11.2
in non-female-dominated occupations	113	19.3	15.9
Part-time employed	111	13.6	15.8
in female-dominated occupations	74	11.9	13.4
in non-female-dominated occupations	37	16.9	20.3
All women	321	15.5	14.8
Men			
Full-time employed	163	20.3	15.7
All men	171	20.4	15.8

Source: Hanson and Pratt, 1990

Table 4.2 Transportation mode for journey to work

Mode	Women (%) (n = 339)	Men (%) (n = 189)
Drive alone	78.5	85.7
Carpool or drive with other	10.3	7.4
Bus	1.2	1.6
Walk	3.8	2.1
Work at home	6.2	3.2

Source: Hanson and Pratt, 1990

it obscures important differences among women. In their analysis of the 1980 PUMS data for the New York metropolitan area, for example, McLafferty and Preston (1991) showed that the commutes of Latina and black women service sector workers were as long as Latino and black men's commutes and were significantly longer than the work trips of white women. In interpreting the Worcester findings, therefore, it is important to bear in mind the fact that the representative sample on which the findings are based is, like the Worcester population, overwhelmingly (90 percent) white.

Although studies of women's and men's travel in the 1960s and 1970s revealed distinct gender differences in access to and use of the different modes of travel (Giuliano 1979, Hanson and Johnston 1985), the gender difference in journey-to-work travel time in Worcester does not reflect a difference in the mode of transportation used (see Table 4.2). The overwhelming choice for both women and men is the "drive alone" mode, which accounts for roughly four-fifths of each group's work trips. While women are more likely than men to drive to work with others, to walk, or to work at home, these differences are small enough to have occurred by chance.

Comparing the work-trip travel times of women in the three different occupational groups exposes a geographical dimension to occupational segregation: the job locations of women working in female-dominated occupations are significantly closer to home than are those of other women, and this difference remains when employment status (working full time versus part time) is held constant. Full-time working women in female-dominated occupations had commutes that averaged 13.3 minutes, compared to commutes of 19.3 minutes on average for other women ($p < .01$). Women in female-dominated occupations who worked part time had the shortest work trips of all groups; their average travel time was only 11.9 minutes, whereas part-time female workers in other occupations averaged 16.9 minutes ($p < .10$). In short, the labor market areas of women working in gender-typical occupations are extremely

Table 4.3 Reasons for working within 10 minutes of home

| | *Percentage of group giving reason* | | | |
| | *Women* | *Men* | *Women in female-dominated occupations* | *Women in gender-integrated and male dominated occupations* |
Reason	*(n = 166)*	*(n = 69)*	*(n = 99)*	*(n = 67)*
Workplace has to be within walking distance/no car regularly available	12.0	2.8	17.2	4.5
Don't like driving	9.6	8.7	9.1	10.4
Don't like to drive in bad weather	6.6	1.4	9.1	3.0
Want to be able to get home quickly for kids, emergencies, family reasons	29.5	2.8	32.3	15.4
Don't like to spend time commuting	22.3	33.3	20.2	25.4

Source: Hanson and Pratt, 1990
Note: These reasons were given in response to question, "Why do you work so close to home?" and were post-coded; only the most frequent responses are presented

small relative not only to men's labor markets but also to those of other women.[3]

Although we reserve for later chapters a full discussion of the reasons for these patterns, we note here some insights gleaned from asking people with commutes of less than 10 minutes why they worked so close to home. The answers disclosed not only large differences between women and men but also distinct differences between women in gender-typical and women in gender-atypical occupations (see Table 4.3). The reason most frequently given by women for working so close to home (cited by 29.5 percent of women with short commutes) had to do with wanting to be able to get home quickly to tend to children or to respond to household emergencies. This was more often a factor for women in female-dominated occupations than for other employed women. Only two men cited family factors as reasons for working within 10 minutes of home; in contrast to women, men were much more likely to explain their very short worktrips by saying they simply did not like to spend time commuting. Lack of dependable access to transportation was another reason mentioned far more frequently by women than by men, and more frequently still by women in female-dominated jobs than by women in other lines of work.

Studying people's actual work trip times provides a one-dimensional index of their labor market geographies. Probing their willingness to travel farther than they now do extends that index and is a window on potential labor market areas. We asked people the maximum time they would be willing to travel to work,[4] and a pattern identical to that describing actual travel times emerges. Men expressed a willingness to spend much more time on the journey to work (about 40 minutes on average) than women (about 27 minutes)(p = .000). Similarly, the maximum time that women in female-dominated occupations are willing to travel is significantly less (25 minutes on average) than that given by women in other lines of work (about 30 minutes) (p < .01).

Clearly women are more sensitive than men are to job location (in terms of the travel time from home), and women in female-dominated jobs are more sensitive than are other women. Women are also more sensitive to another dimension of job location – its spatial stability, its predictability – in that women are more likely than men to work in jobs that have a single, fixed work location. About one-fifth of the men but only one-tenth of the women in our sample had jobs that entailed more than one work site. Most of these people (with jobs like traveling salesperson, telephone line installer, truck driver, visiting nurse, various kinds of construction worker) had a central check-in place they visited almost daily, and this central point was taken as their work location in calculations of home–work separation. But 7 percent of the men and 3 percent of the women had jobs that lacked any stable locational base;[5] these included occupations such as private household cleaner, house painter, property manager, and managers of marketing, advertising, and public relations.

The fundamental gender difference in travel time to work is robust across various household configurations and age cohorts and across most social class groups (Table 4.4). Whether single or married, men travel farther afield to work than do their female counterparts. In line with the findings of other studies (Johnston-Anumonwo 1992; Gordon et al. 1989), the gender difference in work-trip length remains when we control for the presence or absence of children. Although the gender travel-time gap is smaller for those with no children at home, the waxing or waning of active parenthood does not appear significantly to affect women's propensity to work closer to home than their male counterparts.

In Chapter 3 we noted that the Worcester evidence does not encourage the optimistic view that younger women's labor market position is more advantaged than is that of older women. Occupational segregation by sex in Worcester shows no sign of weakening among women under age 40. Similarly, the gender difference in travel time to

Table 4.4 Gender differences in travel time to work

	Women		Men		
	n	Travel time	n	Travel time	p-value
Marital status					
Single, divorced, widowed	63	15.2	37	20.7	.08
Married	259	15.6	137	20.3	.01
Parenthood					
≥ 1 child under age 18 at home	225	14.5	105	20.3	.01
≥ 1 child under age 6 at home	91	15.3	46	21.8	.02
no children at home	100	17.6	69	20.5	.10
Age					
< 40	171	15.8	96	21.3	.01
≥ 40	154	15.1	78	19.2	.04
Social class					
Unskilled manual	28	12.6	35	19.8	.02
Unskilled nonmanual	108	11.9	13	17.7	.10
Skilled manual	20	18.1	41	17.9	n.s.
Skilled nonmanual	108	17.1	40	25.0	.02
Professional/managerial	59	19.6	45	19.8	n.s.
Income ($)					
< 12,000	116	11.2	9	11.1	n.s.
12–19,999	91	15.6	19	19.3	n.s.
20–29,999	74	19.1	55	19.4	n.s.
30–39,999	19	20.3	34	19.3	n.s.
> 40,000	9	21.7	49	24.4	n.s.

Source: Personal interviews, Worcester MSA, 1987
Note: Average travel times in minutes. The *p*-value is from the gender contrast of mean travel times. n.s. indicates that the difference between the two groups is not statistically significant, in other words it is small enough to have occurred by chance

work is no less among younger workers (under age 40) than it is among older workers (Table 4.4). Comparisons of the commuting times of women and men in the different occupational classes show that women's work trips are significantly shorter in three of the five groups (Table 4.4). While the gender difference in commuting travel times remains substantial among unskilled workers (both manual and non-manual) as well as among skilled nonmanual workers, it does not hold for those in skilled manual or professional/managerial jobs. The disappearance of the gender difference in these latter two occupational classes largely reflects the fact that these women spend more time traveling to work than do other women; that is, the disappearance is not due to a shortening of men's trips.

Contrary to other recent findings (Rosenbloom and Burns 1993; Hanson and Johnston 1985), Worcester women and men earning approximately the same income have work trips that are approximately the same length. We find, therefore, an association between income and travel time to work, with lowest-income workers traveling the least amount of time. What attracts attention in Table 4.4 is the gender wage gap indicated in the numbers of women and men in each income category: the number of women drops off rapidly with increasing income whereas the number of men does not.

Women have lower incomes, and lower-income workers have shorter work trips. Does this mean women's work trips are shorter than men's because their incomes are lower? This is doubtless the case for some women, but the qualitative information from the interviews indicates that for some women the causal arrow is reversed: their incomes are lower because their work trips are shorter.

In addition to the persistent gender gap in work-trip travel times, significant differences are also evident among groups of women beyond those differences we have already described having to do with occupational segregation. One significant similarity, as opposed to a significant difference, that seems worthy of attention is that younger women (those under age 40) do not, on average, travel farther to work than do older women (Table 4.4), and this holds true when we control for part-time or full-time employment status. It is also interesting that married women work no closer to home, on average, than do single women (Table 4.4). Women who work in professional/managerial and skilled occupations do, however, have distinctly longer work trips than women who hold jobs in occupations classified as unskilled (Table 4.4) ($p < .01$). These disparities among women working in different types of jobs remain when only part-time or full-time workers are considered and are consistent with the pattern described earlier for women in gender-typical and gender-atypical occupations.

We have described both a durable gender gap and significant differences among women in journey-to-work travel times. Both are visible in maps of the commuting distances of the women and men living in each of the four study areas (Figure 4.2).[6] Women's orbits are distinctly smaller than men's, but the median distance travelled to work by women living in middle-class Westborough (6 miles) is considerably longer than are those traversed by women living in Main South (1.3 miles), Upper Burncoat (3 miles), or the Blackstone Valley (3 miles). Particularly striking is the large number of men (but the very few women) who work outside the Valley; mindful of the fact that better-paying jobs are to be had outside the Blackstone Valley, about two-thirds of the Valley's work force was employed elsewhere by the mid-1980s, according to an

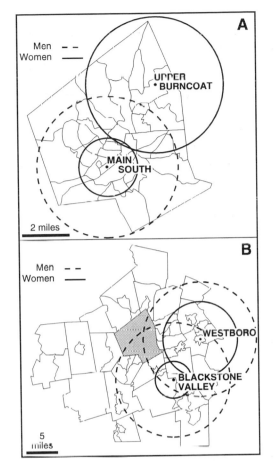

Figure 4.2 Median journey-to-work distances of men and women living in four
local areas in Worcester, Massachusetts; (A) City of Worcester, (B) Worcester MSA
Source: Pratt and Hanson 1994; personal interviews, Worcester MSA 1987

estimate made by the head of the Blackstone Valley's Regional
Development Corporation (Jones-D'Agostino 1987).

IMPLICATIONS AND REFLECTIONS

Analysis of work-trip length shows clearly that women work closer to
home than do men and that women in female-dominated occupations
work closer to home than do other women. Accordingly, the spatial
extent of women's labor markets is smaller than men's, and women in
gender-typical occupations have labor markets that are smaller than
those of other women. These findings raise several questions about the

103

implications for understanding gendered labor markets. First, how do we interpret these findings; is a short worktrip to be considered "good" or "bad"? Second, what role does the mobility of the residential location play in generating the work-trip lengths we observe? Third, how might the spatial distribution of employment opportunities contribute to these findings? Finally, to what extent might women's shorter work trips be related to the gendering of work? Before examining maps of employment opportunities, we briefly consider each of these questions in turn.

First, how do we interpret these findings? Is a short commute to be valued as a time saver (as clearly it is by many of the men who work close to home) or condemned as a curb on a person's employment opportunities? As we shall see in the following chapters, asking whether women's short commutes are the result of "choice" or "constraint" is far too simplistic; posing the question in terms of a clear-cut "either/or" yields few useful insights. Sometimes women clearly choose a job close to home as part of a carefully conceived time-management strategy that enables them to care for their family while earning a living. At other times, the constraints imposed by the paucity of available alternatives clearly overwhelm any sense of "choice." Most often the decisions women make about where to work – the decisions that lie behind their shorter work trips – are deliberate choices made within the context of constraints such as unreliable transportation, inadequate child care facilities, and husbands who are not equal partners in homemaking and childrearing. In our view, the "goodness' or "badness" of working close to home depends on how much choice is involved and on the wage earned at the end of the worktrip.

The woman from Uxbridge we described earlier in this chapter is a good example. When we interviewed her, she was working in the shipping department of a yarn factory five minutes from her home. Immediately prior to this job she had been working as a warehouse picker 40 minutes away in Westborough but had decided to take a pay cut in order to find a job close to home to be available for her school-aged children. This woman was keenly aware of the wage penalty she was incurring by this decision, "I would prefer to get a better job, but Uxbridge doesn't pay as well as other areas." Her husband, with the same education level as his wife, had left his previous job in Uxbridge to commute to Westborough for higher pay, but she had explicitly chosen a job that offered low wages and proximity to her family over one that entailed sacrificing availability to her children for higher wages. What is important here for any theoretical understanding of gendered employment geographies is the clear sense that low wages were the consequence, not the cause, of this woman's very short commute. In Worcester as elsewhere travel time to work is positively related to income, but it is misleading, as is often implied in the literature (e.g.,

Madden 1981: 193; Wekerle and Rutherford 1989: 154), to infer that this correlation means that women's short commutes are the *result* of having low-wage jobs, that people with poorly paying jobs work closer to home because they have no incentive to look for work farther afield.

Our finding about the differential friction of distance for women and men, and for different groups of women, prompts a second question that must be answered if we are to understand the implications of these patterns for people's access to jobs. The workplace is simply one end of the journey to work; travel time to employment depends on the location of home as well as of work. How much freedom do people have to move their place of residence to accommodate a particular job location? Where such freedom is abundant, short work trips could simply reflect the flexibility of residential location together with a desire to minimize travel. Such geographic maneuvering can become a geographic juggling act however when one residential location has to accommodate access to more than one work place. And this is the norm: 57 percent of the households in our survey had more than one wage earner; 80 percent of the respondents were living with long-term partners and in two-thirds of these married or cohabiting couples both partners were in the paid labor force.

How mobile is the home location in these circumstances? In general, we found that most households did not choose their residential location to accommodate a job location: only 7 percent of the women and 37 percent of the men we interviewed said that they had gotten their present job first and then found their current residence (these percentages remain unchanged when only married women and men are considered). In short, for only a very few women does the possibility exist that short commutes are the outcome of a strategy to locate home close to work. For six out of ten men and for nine out of ten women the residential location is essentially fixed prior to their finding a job and is not easily shifted in response to a job location.

Coupled with women's short commutes, residential immobility raises a third question: how does the nature of job opportunities vary over space? What does the employment map look like? In other words, how much does it matter *where* that relatively immobile home place is located? If every type of job were scattered evenly across the urban landscape, then any residential location on that landscape would offer the same job accessibility to all people willing to travel a given distance. If, however, certain types of work are clustered in certain parts of the city, then a given residential location will offer better access to some lines of work and worse access to others. As we describe in detail in the following section, jobs are anything but uniformly distributed: they are distinctly unevenly spread throughout the metropolitan area. Moreover, they are divided – at a fine spatial scale – along gender lines, with certain districts rich in female-dominated jobs and others offering predominately jobs in male-dominated occupations.

These facts, piled one on top of the other – women's short travel times to work, their residential fixedness, the lumpiness of the employment map – raise yet another question in our quest for understanding gendered labor markets: to what extent and how might geography contribute to the gendering of jobs? Might employers, cognizant of the stickiness of women's residential locations and aware of gender differences in the friction of distance, actually be partially responsible for creating spatial clusters of female-dominated jobs? While we explore these questions in greater detail in Chapters 6 and 7, we simply want to raise here – in the context of the spatial patterns that are the focus of this chapter – the point that whereas we have taken the categories of, for example, part-time work or female-dominated occupations somewhat unproblematically "off the shelf" and looked at the commuting patterns of women in these groups, the categories may themselves be shaped, in part, by the patterns we find.

Part-time work may, for example, be created as women's work (Beechey and Perkins 1987) in locations that are close to women's homes. Conversely, in many places city and state ordinances in force throughout the 1960s prohibited women from bartending jobs in part because the night-time journey to work was deemed too dangerous for women (Cobble 1991). As Henrietta Moore (1988: 36) reminds us, the cultural construction of gender emerges from "the actual social relations in which gendered individuals live," from the competing claims that women and men make on each other in specific circumstances. The experiences of the young woman in Uxbridge we described earlier ground us in the recognition that distance, place, context – in short, geography – are implicated in those social relations and integral to the construction of gender.

WHERE THE JOBS ARE

The location of employment can be viewed as the spatial expression of the demand for labor. What does this demand picture look like? We consider two questions in assessing the employment maps of Worcester. One has emerged from studies of gender differences in the separation between home and work: are women's work trips shorter than men's, and trips to female-dominated jobs shorter still, because jobs employing women are more evenly spread across urban space? If women's work places were both abundant and uniformly distributed across the map, then this spatial arrangement could help explain the journey-to-work patterns we have observed. The first question we consider is, then, whether the spatial distribution of women's jobs is more uniform than that of men's. The second question flows out the confluence of two streams of thought, one springing from a concern with the gender division of labor, the other from a concern with a spatial division of labor: is the gender division of labor, so well documented within the U.S. and Canada, visible on the ground; is it manifest in a spatial division of labor within metropolitan areas? Analysis of the data from the census journey-to-work files allowed us to answer these two questions.

To find out whether women's jobs were distributed more evenly than men's jobs over the urban landscape we calculated, separately for women and for men, the proportion of the metropolitan-wide employment base that was located in each census tract (see Figures 4.3 and 4.4). If women's jobs were scattered more evenly, we would expect Figures 4.3 and 4.4 to be quite different; in fact they are remarkably similar. It is clear from these maps that neither women's nor men's jobs are dispersed uniformly across all census tracts; instead, employment tends to be concentrated in certain census tracts whereas other tracts tend to be primarily residential. In those tracts that are sites of employment, the proportion of metropolitan-wide male employment is not dramatically different from the proportion of metropolitan-wide female employment. In sum, these maps do not support the idea that the reason women's work trips are shorter than men's is that women's jobs are distributed more uniformly and therefore on average reached more easily than men's jobs are.

As we described in Chapter 3, the gender division of labor – the sorting of women and men into different occupations – is starkly evident in the Worcester labor force. The second question we address in considering the location of jobs is whether or not this gender division of labor is evident on the employment map when we look at the distribution of employment for women and for men within small areas. The idea that the gender division of labor has a spatial expression has received some play in descriptions of the global economy; spatial clusters of predominately female employment are created when firms seek low-wage female labor in particular places, for example along the U.S./Mexico border (Christopherson 1983) or in South Korea (Cho 1985). Similarly, Massey (1984) has described a gendered spatial division of labor at the regional scale within the United Kingdom. Although the extreme brevity of women's work trips suggests that such gendered spatial divisions might also prevail within cities, the lack of gender-specific employment data at the census tract level has blocked empirical analysis of the spatial dimension of the gender division of labor at a submetropolitan scale.

We pursued this question by calculating the proportion of jobs in each census tract that were held by women and by men (Hanson and Pratt 1988b). Recognizing that employment tends to be concentrated in

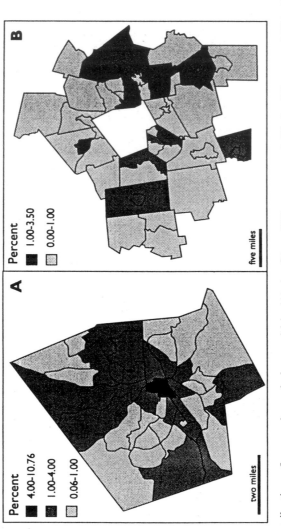

Figure 4.3 Distribution of women's workplaces within (A) the City of Worcester and (B) the Worcester MSA. The mapped values are for 1980 and show the number of women with workplaces in a given tract divided by the number of all employed women in the MSA

Source: Hanson and Pratt 1991; special runs on the 1980 Worcester, Massachusetts MSA census journey-to-work file

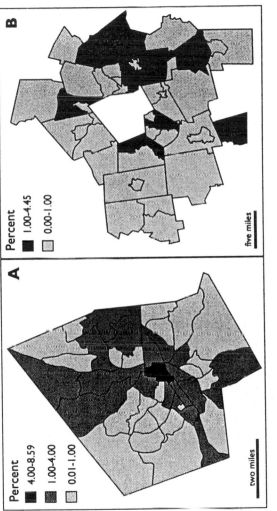

Figure 4.4 Distribution of men's workplaces within (A) the City of Worcester and (B) the Worcester MSA. The mapped values are for 1980 and show the number of men with workplaces in a given tract divided by the number of all employed men in the MSA

Source: Hanson and Pratt 1991; special runs on the 1980 Worcester, Massachusetts MSA census journey-to-work file

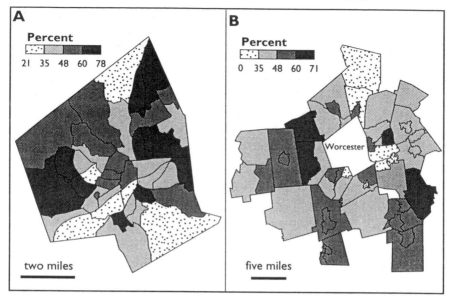

Figure 4.5 Proportion of each census tract's employment held by women: (A) City of Worcester; (B) Worcester MSA
Source: Hanson and Pratt 1988b, 1990; special runs of the 1980 Worcester, Massachusetts MSA census journey-to-work file

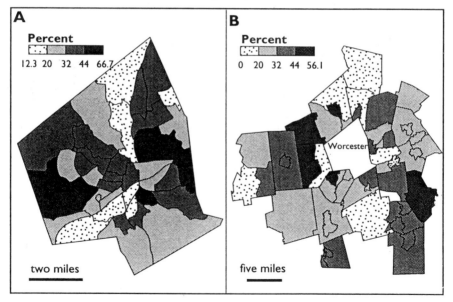

Figure 4.6 Proportion of employment in each census tract that is in female-dominated occupations. (A) City of Worcester; (B) Worcester MSA
Source: Hanson and Pratt 1988b, 1990; special runs of the 1980 Worcester, Massachusetts MSA census journey-to-work file

certain parts of the city, we asked whether each census tract offered the same proportion of its jobs to women and to men. The answer is decidedly no! As shown in Figure 4.5, the people working in certain tracts are predominately women, and in other tracts they are predominately men. As 43 percent of the entire Worcester MSA labor force was female in 1980, the expected value for percent female within each tract is 43. Yet in 15 census tracts more than 60 percent of the workers are women and in 13 census tracts more than 65 percent of the workers are men. Within each census tract the job opportunities for women and for men are decidedly not the same.

This pattern is replicated in the maps showing employment in female-dominated occupations and male-dominated occupations. Although only 29 percent of the total metropolitan workforce is employed in female-dominated occupations, there are 11 tracts in which more than 44 percent of the jobs are in female-typed work. Similarly, in 14 tracts, more than 54 percent of the jobs are in male-dominated occupations although "only" 44 percent of the total workforce is employed in these occupations. Clearly, female-dominated jobs and male-dominated jobs are located in different census tracts. Visual comparison of Figures 4.6 and 4.7 indicates that the two maps are practically the inverse of each

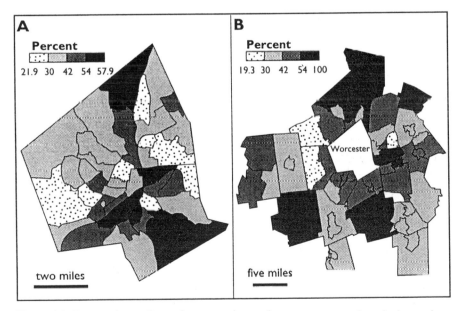

Figure 4.7 Proportion of employment in each census tract that is in male-dominated occupations, (A) City of Worcester; (B) Worcester MSA
Source: Hanson and Pratt 1988b, 1990; special runs of the 1980 Worcester, Massachusetts MSA census journey-to-work file

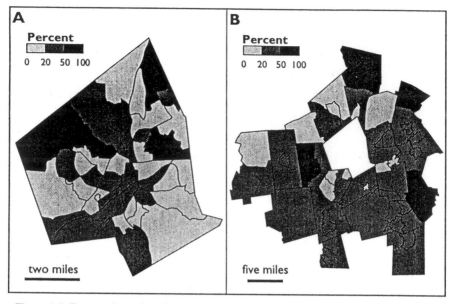

Figure 4.8 Proportion of each tract's manufacturing employment held by men in
male-dominated occupations, (A) City of Worcester; (B) Worcester MSA
Source: Hanson and Pratt 1988b; special runs of the 1980 Worcester, Massachusetts MSA
census journey-to-work file

other, suggesting that the gender division of labor does indeed have a
spatial expression at a fine geographic scale.[7]

Are these maps simply a cartographic translation of the fact that
women and men tend to work not only in different occupations but
also in different industries (see, for example, Tables 3.1 and 3.2, pp. 58
and 59)? In light of the fact that certain industries tend to be concentrated in certain areas of the city (for example, manufacturing in some
districts, hospitals or retailing in others), do these maps simply mirror, in
two-dimensional space, the gender schisms by industry that cleave the
labor force? To some extent they do. Some of the tracts with high female
employment, for example, house large hospitals; others are the sites of
large shopping facilities that employ many women as sales clerks and
cashiers. When we map male and female employment within a given
industry, however, the gendered spatial divisions remain visible. The
maps of jobs in manufacturing (Figures 4.8 and 4.9) and in health,
education, and welfare (Figures 4.10 and 4.11) show clearly that even
when women and men work in the same industry, their jobs are located
in different parts of the urban area, suggesting that women and men
work in different types of firms or establishments.[8]

Taken together these maps describe a highly variegated employment

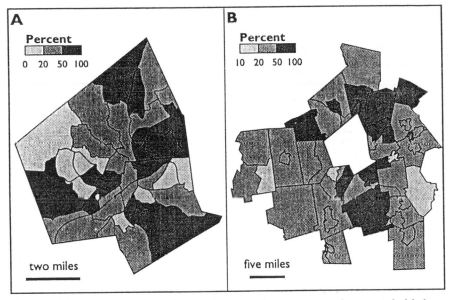

Figure 4.9 Proportion of each tract's manufacturing employment held by women in female-dominated and gender-integrated occupations, (A) City of Worcester; (B) Worcester MSA

Source: Hanson and Pratt 1988b; special runs of the 1980 Worcester, Massachusetts MSA census journey-to-work file

landscape, a landscape of uneven employment distributions for both women and men. What this unevenness suggests is that residential location is crucial in defining access to particular types of work.[9] These maps also show that the spatial patterns of demand for women's work and men's work are distinctly different, suggesting that, even if women and men traveled on average the same distance or time to work, a particular residential location is unlikely to offer the same job accessibility for a woman and a man. Because women's work trips are in fact considerably shorter than men's, residential location is likely to be more important for women in defining accessible employment opportunities, especially so for women living in suburban locations like Westborough, where the overall density of employment opportunities tends to be lower. We turn now to consider what effects access to different types of employment has on occupational segregation.

CONTEXT, GEOGRAPHY, AND OCCUPATIONAL SEGREGATION

It is perhaps obvious that the type of work people do depends to some extent on what is available in a particular time and place. The familiar

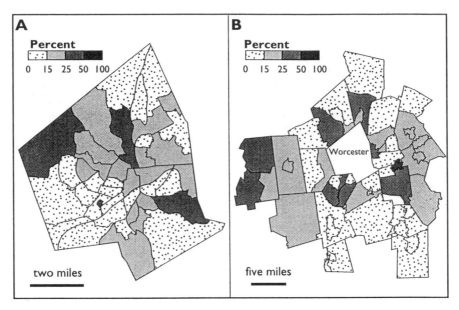

Figure 4.10 Proportion of each tract's health, education, and welfare employment held by men in male-dominated occupations, (A) City of Worcester; (B)
Worcester MSA

Source: Hanson and Pratt 1990; special runs of the 1980 Worcester, Massachusetts MSA
census journey-to-work file

example of making buggy whips comes to mind. Jerry Jacobs (1989) has
argued that the persistence of gender-based segregation in the work
place is largely the outcome of a socialization process that conditions
the aspirations and expectations of girls. Specifically, this socialization
process shapes the opportunities that girls see as being available to them:
"opportunity generates interest" (158). According to Jacobs, women
enter male-dominated occupations as the social controls that have kept
them out of such occupations are weakened; as women see more
opportunities for them opening up in male-dominated work, women's
aspirations change, prompting even more of them to enter gender-
atypical lines of work. In contrast to the human capital explanation
for occupational segregation, Jacobs's argument emphasizes that
women's employment decisions are short term (not some long-range
calculus that attempts to maximize life-long earnings) and are extremely
responsive to the availability of opportunities. More than 20 years ago
Valerie Oppenheimer (1973) made a similar demand-based argument
for understanding female labor force participation patterns in the U.S.
She saw the post-World War II increase in women's entry into paid work
largely as the result of economic changes that led to increased demand

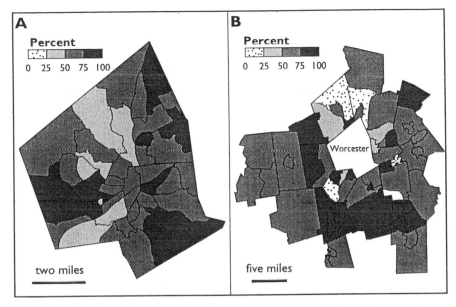

Figure 4.11 Proportion of each tract's health, education, and welfare employment held by women in female-dominated and gender-integrated occupations,
(A) City of Worcester; (B) Worcester MSA
Source: Hanson and Pratt 1990; special runs of the 1980 Worcester, Massachusetts MSA
census journey-to-work file

for women in particular types of work (such as clerical positions)
considered appropriate for women.

Both Jacobs and Oppenheimer recognized the importance of context
in attempting to understand occupational segregation and women's
entry into the paid labor force. But they have conceptualized and
operationalized context almost entirely in temporal or historical terms.
Their geographic point of reference is the U.S. as a whole. In our view,
context needs to be specified not only historically but also geographically.
Jacobs, for example, suggests that women in different social contexts (in
which different social controls prevail) will receive different information
about the job opportunities that are available to them. Certainly the
information people absorb about employment opportunities is likely to
vary not only over time, as a function of the prevailing national social
context (which is what Jacobs has in mind), but also from place to place.
Here the question of geographic scale once again is crucial.

As long ago as 1925, Gwendolyn Salisbury Hughes, in her study of
women and work in Philadelphia, noticed that mothers' labor force
participation and the types of paid work that mothers were engaged in
varied by neighborhood within Philadelphia because the opportunities

115

for women's employment differed from district to district. She also noted that in the southeast district, where the overall poverty of the population led many women to seek paid work but where no local industry had yet located, the women traveled longer distances to their workplaces and were involved in a greater diversity of occupations than were women from other districts (Hughes 1925).

More recently, a number of quantitative studies of large data sets have specified context geographically and examined the impact of local context on women's occupational structure and labor force participation. All of these efforts have taken local context to be the entire metropolitan area (the Metropolitan Statistical Area (MSA) in the U.S. or the Travel to Work Area (TTWA) in the United Kingdom), examining how women's labor market outcomes depend in part on characteristics of the metropolitan area. These studies have shown that, in fact, city size and the industrial mix in a metropolitan area do have an impact on women's labor force participation rates (Ward and Dale 1992) and on the types of work that women do (Abrahamson and Sigelman 1987; Jones and Rosenfeld 1989).[10] Places where most of the jobs have been in traditionally male-typed work (mining, steel making, ship building) have had low rates of female participation in the paid labor force (Massey 1984), and places with many job opportunities in female-typed work (assembly of garments, electronics assembly) have higher proportions of women in the work force.

As noted in Chapter 1, Walby and Bagguley (1990) have criticized those who analyze regional variations in female employment rates as reflections of local industrial structure. They charge that framing explanations of women's employment patterns in this way implies that occupations are rigidly sex typed, whereas the sex composition of occupations is known not only to vary from place to place but also to change over time. Although the gender composition of occupations and industries does indeed change and although this change can be quite rapid, sometimes occurring within a decade (Fine 1990; Reskin and Roos 1990), evidence also suggests that the types of jobs available in the local labor market do affect employment outcomes for women, especially in the short run. We do not argue here that occupational segregation is completely dependent on local labor market opportunities but instead that local context does play a role in labor market outcomes especially for some groups of women. Anecdotal evidence of the extent to which this is true hit us many times in the Worcester interviews when we posed the question, "Could you trace the circumstances and decisions that have led you to do the kind of work you are doing?" For many, many women, the type of work they were doing was not the outcome of any reasoned decision; rather, the reason given was simply "because it was there."

The likely importance of local context for women's labor force decisions has been widely recognized (Moseley and Darby 1978; Dex

116

1987; Clark and Whiteman 1983; Stoltzenberg and Waite 1984) but has proven difficult to study empirically. The problem lies in capturing the local employment context in appropriate measures at an appropriate geographical scale. It is clear from our earlier analysis of work-trip travel times that that scale must be smaller than the entire metropolitan area. We use the employment maps of job opportunities in the Worcester area to assess the impact of employment context at the census tract level on occupational segregation. We focus on the question of whether the types of jobs available locally affect the occupational segregation of women.

Answering this question requires tackling a number of important measurement issues: we define job type in terms of the gender typing of occupations (female-dominated, male-dominated, gender-integrated) and consider whether or not local context affects the likelihood that a woman is in a female-dominated occupation. We determine the "local availability" of female-dominated jobs by measuring the density of jobs in female-dominated occupations within a certain distance of a woman's home. To reflect the fact that the distance women are willing to travel to work varies with city versus suburban residence and with full-time versus part-time employment, the distance used in each case depends on whether the woman lived in the city or a suburb and whether she worked full or part time.

For each woman in the sample, then, we calculated a number of variables measuring the nature and density of locally available employment opportunities, variables such as the density of all jobs in female-dominated occupations and the density of part-time jobs in female-dominated occupations. The variables were calculated as follows. Each woman's home location was placed on the base map of the type of employment in question (e.g., the map of jobs in female-dominated occupations; see Figure 4.12), and a circle was inscribed around her home on the map. Because average commute lengths vary with urban density and with a person's employment status, the radius of the circle depends on the woman's residential location (city or suburb) and employment status (full time or part time). If, for example, she was living in a suburb and working part time at the time of the interview, the radius used to define her locally available employment opportunities would be the median distance traveled to work by all suburban part-time employed women in the sample.[11] The geographic information system, IDRISI (Eastman 1992), enabled the calculation of the number and density of jobs within the radius, not a simple matter when the circle encompasses several census tracts (see Figure 4.12).

As noted in the previous chapter, several non-spatial factors are strongly related to the gender composition of a woman's occupation. Having at least one preschool child at home, working part time, and having no more than a high school education all increase the likelihood that a woman will work in a female-dominated occupation. We therefore

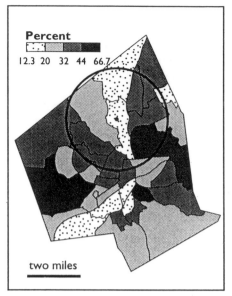

Figure 4.12 Dot represents residential location. For each woman the number of employment opportunities of a particular kind (here, in female-dominated occupations) within a certain radius of home (here 1.86 miles because the woman lives in the city and works part time) is calculated by determining what proportion of each census tract falls within the journey-to-work radius

divided the sample along these lines into six groups[12] and for each group used discriminant analysis to assess whether the variables measuring local employment context were effective at discriminating between women who worked in female-dominated occupations and women who worked in other types of jobs.

The results show that local employment context matters most for women who have preschool children at home and who work part time. For these women, having a large number of part-time jobs in female-dominated occupations in the vicinity of home increases the likelihood that they will have jobs in gender-typical occupations. Interestingly this is particularly the case for more highly educated women (those who have gone beyond high school). Among women with some higher education, then, those who work part-time in gender-typical occupations tend to live in areas where the density of jobs in part-time, female-dominated lines of work is higher than it is for similar women[13] who work in male-dominated or gender-integrated occupations. For the other groups of women (e.g., those working full time or with no young children at home), the local context variables did not effectively discriminate between women in female-dominated occupations as opposed to those in other types of jobs.[14]

These results make sense in that local employment context is likely to have the biggest impact on the type of job a person has when that person has decided, for whatever reason, to find a job very close to home. For a variety of reasons, which we explore in greater depth in the next chapter, such people are often women with young children.[15] In the case of well-educated women, the decision to work nearby forecloses other job opportunities that might be open to them but not to those with less formal education. Women with only a high school education or less are more likely than women with some higher education to find themselves in female-dominated occupations regardless of the local employment context. This analysis suggests, in sum, that the kind of jobs available locally can be associated with occupational segregation, but only for a segment of the female work force: mothers of young children.

CONCLUSION

To say that the type of job a women can find depends in part on "what's available in this geographic area" (Towne 1979) might seem obvious. Yet it is not. Few, if any, women (or men) search for jobs throughout an entire metropolitan area, and the geographic size of the labor market (as measured by commute times) is different for different groups of women. Grasping what the appropriate geographic area is and identifying the nature of the job opportunities therein has proven elusive because the standard geographic units for which employment data are released are simply too large.

Clearly for most women and men, occupying the same home base does not mean sharing access to the same labor market area. This is not only because of labor market segmentation, but also because most women and men have different commuting ranges. Moreover, the employment landscape is highly variegated in terms of the gender composition of the available jobs: some parts of the metropolitan area are rich in female-dominated work, others in male-dominated jobs. For people looking for work close to home (most of whom are women), the residential location on this landscape, the location from which the job search begins, importantly defines access to different kinds of work. The local employment context is particularly important for well-educated mothers of young children seeking part-time employment: these women are more likely to find jobs in female-dominated occupations if the local density of part-time female dominated work is relatively high.

5

HOUSEHOLD ARRANGEMENTS AND THE GEOGRAPHY OF EMPLOYMENT

"Distance makes the heart skip for commuter moms: the perils of the modern Pauline" (Lawson 1991): the *New York Times* article that follows makes it clear that commuting complicates and compromises a woman's capacity to mother responsibly. "The phone rang. Barbara Reisman's nightmare was on the line" (C1). The school nurse had called to tell her of her child's injury; sitting in her Manhattan office one hour away from home, Ms. Reisman was able to deliver her child to a doctor only after complicated transactions with her husband and only "just in time" (C1). Ms. Reisman admits that "there are moments when I think I'd be better off if I had a job closer to home" (C1). In the life stories that follow in this particular article, woman after woman, modern Pauline after modern Pauline, tell of the perils of a long commute and the complicated arrangements that they have made – nannies, relying on friends within the neighborhood, compensating neighbors in the evening and on weekends – to offset their daily absence. Their husbands, who undoubtedly also commute, appear to feel none of the strain.

A vaguely threatening and disciplining tone pervades this article: take a high-powered job in distant Manhattan at your (or, worse, your child's) peril! Just as disturbing, however, is the assumption that mothers will and should take primary responsibility for child care; the absence of fathers from this *Times* piece is palpable. Yet this distinct woman-as-primary-caregiver bias no doubt captures a pervasive truth about gender and work in contemporary American society. In this chapter, we turn to examine how such patriarchal assumptions and practices within households affect women's employment. Other studies (e.g., Hartmann 1987) have shown that, despite increases in women's labor force participation, traditional gender relations within the home persist. After studying work arrangements in households in Southampton, England, for example, Pinch and Storey (1992) conclude: "The survey evidence presented . . . would add further weight to those who argue that reported sightings of the so-called 'New Man' have been exaggerated" (11). Studies that report that men are affected by their wives' employment

(Pleck 1985) find that their response comes in the form of increased "availability" to their children, but not in terms of housework or more narrowly defined child care (Morris 1990).

We have looked within a sample of Worcester households to understand how decisions about where and how to live affect women's and men's labor market experiences. We are especially concerned with the geography of these strategies.[1] Our overall impression is similar to that of Pinch and Storey: like them, we have made few sightings of the "New Man" (or modern Paulines), and in the first part of the chapter we tell traditional stories about how household strategies reflect and reproduce patriarchal gender relations; the novelty to these stories lies in highlighting the geography of these practices. In the latter part of the chapter we try to tell new stories, by drawing out the diversity of strategies developed by different types of households.

GEOGRAPHIES OF PATRIARCHAL FAMILIES

Much has been written about the family as a site of patriarchal relations (Barrett and McIntosh 1982).[2] Attention has focused, in particular, on how familial ideologies shape men's and women's visions of possible and appropriate ways of living and how many women's responsibilities for domestic work wed them to the home. When the implications of familial relations for women's labor force participation are considered, typically the focus has been on how familial ideology limits women's wages, job security, and access to a full range of occupations (Barrett and McIntosh 1982); on the ways that temporal constraints imposed by domestic work shape women's employment (e.g., by dictating part-time work, preventing women from attending union meetings); and on the long hours of work experienced by many women relative to their male partners, with attention given to women's double (and, including community work, triple) "days" of work (Hartmann 1987).

Social and economic relations take place in space as well as time; feminist geographers have drawn attention to this by showing how domestic responsibilities effectively restrict women's employment options if they must look for work close to home. Our goal in this section of the chapter is to extend this understanding by showing how a variety of household arrangements lead women into relatively disadvantaged positions in the labor market and by tracing how geography enters into this process.

Rooted in place – not necessarily of her choosing

Social theorists tend to accentuate the mobility of contemporary individuals, households, businesses, capital flows, and cultural practices. While this tendency captures one aspect of contemporary social reality, we place along side it a very different set of spatial relations, by calling attention to the remarkable rootedness of many people's (and especially women's) lives. By focusing on households, rather than individuals, one can also question the assumption – no doubt reflective of class, gender, racial, and neocolonial privilege (hooks 1992; Wolff 1993) – that mobility, when it does occur, is a matter of individual choice.

Worcester stands as possibly an extreme case to make the point about the rootedness of households, although the dearth of comparable data on rootedness makes comparisons with other places difficult. One measure is provided by the U.S. Census of Population and Housing: the proportion of people living in a different residence five years previously. When Worcester is compared to other American cities, it is remarkable for its

Plate 5.1 With three apartments, one on each floor, the three-decker often housed three generations of the same family. Photograph by John Ferrarone

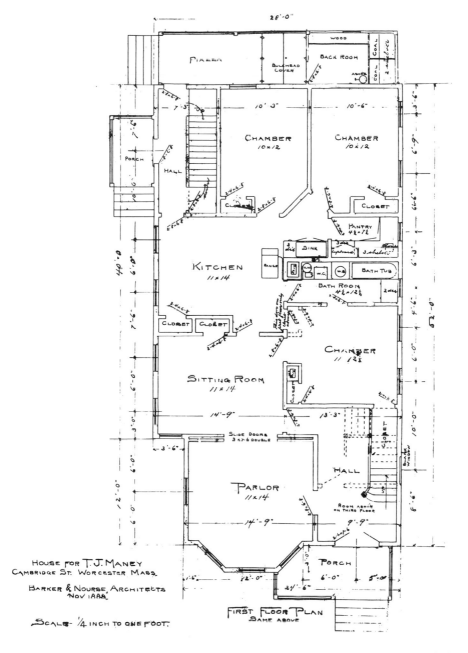

Figure 5.1 The typical three-decker apartment has three bedrooms, a full
kitchen, a sitting room, and a parlour
Source: Spear 1977: 18, 19

low levels of residential mobility. In Worcester only 38.7 percent of the resident population in 1980 had lived in a different house five years before, compared to 47.5 percent for all U.S. metropolitan areas.[3] Reinforcing this point, we found that in our 1987 sample of Worcester households the majority of men (51 percent) and women (55 percent) had lived in Worcester at least 90 percent of their lives. In addition, 13 percent of those who had not themselves grown up in the Worcester area had spouses who had. Among men, this rootedness is especially evident for manual workers: 77 percent of men working in these occupations had lived virtually all of their lives in Worcester, compared to 38 percent of those working as managers or professionals and 31 percent working in skilled nonmanual occupations.[4]

The residential stability of many working-class households may both reflect and account for a housing strategy that is remarkably prevalent among these families, that is the intergenerational transfer of domestic property. About one in five households in which the male household head was employed in a manual job either inherited, bought, or rented a house or apartment from an older family member.[5] One characteristic of the housing stock in Worcester encourages this sharing of housing across generations. Much of the working-class housing stock still takes the form of the three-decker (Plate 5.1), a free-standing house with three generously sized, self-contained units (Figure 5.1), laid out horizontally, one upon the other. Three-deckers were built from the 1860s to 1920s, often for the purpose of housing the extended family within different units of the structure, and they continue to encourage this pattern of residence to this day (Warren 1987).

The implications of this housing strategy are many (Pratt and Hanson 1991): we focus here on its geography and its uneven gender impacts. One effect of inheriting housing is that housing costs are likely lower, and therefore less income is required to sustain the household. Some households take this circumstance as an opportunity to spend fewer hours in paid employment, and it is typically the woman who remains out of the labor force or who opts to work outside of the home only on a part-time basis.[6]

A striking characteristic of inheriting or sharing housing is that its location typically bears little relation to any household member's current employment needs. In numerous cases in which households had inherited domestic property, respondents (both male and female) spontaneously mentioned the restricted employment opportunities within their residential communities, especially if they had inherited a family home in the communities south and west of Worcester, where the regional high-technology boom has had little impact on the local economy. One respondent worked as an operator in a waste water treatment plant in a town southwest of Worcester. He had searched

124

for this job for three years and noted the lack of job opportunities in the area: "It's all factories around here . . . woolen mills. It's a depressed area . . . if I hadn't inherited the house, I wouldn't have been able to afford to own. It limits where you can go, though. I might have moved away if I didn't have a house." A female respondent from another household living in the same town noted that: "If you didn't have any transportation, you would be limited because there are only certain jobs in Webster. It's a factory town mainly. Everything else is retail." She worked part time as a teacher's aide in a school just 4 minutes from her home, but her husband commuted 45 minutes to a job as a computer electronics technician. They would prefer to live in some other town, but housing affordability problems led them to accept her father's offer to move in with him when he inherited the house (they planned to purchase it from him eventually; her sister lived in the apartment upstairs). Asked whether there were conflicts between home and paid employment, she answered: "No, as long as my husband doesn't mind traveling so far."

The last case points to an important discrepancy in the employment opportunities for most men and women who inherit or share domestic property, and to ways that inheriting housing affects men and women differently. Women typically work close to home and are therefore highly dependent upon local employment opportunities. In ten households that inherited domestic property (about one out of six), an individual member compensates for local employment restrictions by commuting a long distance to paid employment (between 30 minutes to an hour). In only one of these households is it a woman who commutes a long distance to overcome the restrictions in her local labor market – and she is single.

The tendency for women to reside within labor markets that may bear no relation to their skills or employment needs is in fact quite typical and goes far beyond households that obtained housing through family members. As we mentioned in Chapter 4, personal histories of the sequence of decision making about jobs and residential location among all households in Worcester substantiate this claim. Describing the order of the decisions with reference to their present jobs, 93 percent of women said they had searched for their present job from a fixed residential location while this was true for "only" 63 percent of the men. This pattern also holds for the two jobs people had held before the current one.

Pursuing the theme of differences, however, the ordering of residence and work varies for both men and women, depending on other social characteristics. First, women who have never married are far less likely than are other women to search for a job from a fixed residential location and, with 32 percent of never-married women finding their residence after their job, their profile looks much like the profile of men. Second, for both men and women, residential rootedness is far less common among people with professional and managerial jobs. In middle-class Westborough, the tendency for men to fit residential location around a job is particularly noteworthy, with only 36 percent finding their current job from a fixed residential location. In the Blackstone Valley, on the other hand, three out of every four men find their present job from an already established home base. Though the differences are not nearly as marked among varying groups of women, those in professional and managerial occupations are indeed more likely to find a residence after finding a job; this was the case for 17 percent of married women in this occupational grouping.[7]

What is particularly interesting, however, is how differently women and men respond to residential rootedness. Men who are rooted in residential space tend to compensate for this by commuting long distances to work; by contrast, women with fixed residential locations find jobs very close to home. Blackstone Valley men, for example, tend to travel very long distances to work (about 35 minutes, compared to the average 25 minutes traveled by Westborough men). On the other hand, married, nonprofessional women, who are especially rooted in residential space, tend to have the shortest commutes.

Responses to other questions support the interpretation that many men have freedom in tailoring their residential locations to job requirements while most women accommodate their job search to an existing residential location.[8] In couple-headed households, most men (60 percent) thought that convenience to their own job was an important consideration in choosing their current residential location, while a much smaller proportion (44 percent) thought that convenience to their wife's job was an important factor.[9] Furthermore, both men and women agree that the man's job is far more likely than is the woman's to lead the household to change its residential location; 59 percent of men indicated that they would be ready to relocate outside of the Worcester area for a better job (only 23 percent thought that the household would relocate if their wife could get a better job). The responses of women interviewed mirrored those of the men.[10] Moreover, among women the type of work one has is clearly related to the willingness to relocate outside the Worcester area for a better job, with married women in female-dominated occupations less likely than other employed women to voice a willingness to move to improve their own employment.[11] These variations among women, although statistically significant, pale in comparison to the much larger gender difference.[12]

Actual histories of respondents over the last ten years confirm these hypothetical conjectures. Households have been more likely to change residential location for the sake of the man's job; 15 percent of female respondents but only 2.5 percent of men had actually changed residence because of a change in the partner's job location. Furthermore, residential moves that are prompted by a change in a partner's job location have an impact on women's job histories but not on men's. Roughly 10 percent of the women in our sample had had to leave a job in order to accommodate a change in their partner's job locations; not a single man had left a job for this reason.

Gender differences in the ordering of the residence and employment location decisions suggest that most women are doubly constrained. Most women must select a job that is close to home, and yet their constrained set of employment opportunities has not been chosen with their employment aspirations or needs in mind. Among married women, employment-related residential mobility is almost always tied to their partner's job and often disrupts their own employment.

Parenting and the geography of paid employment

Families develop all sorts of strategies for combining paid employment with the care of children. It is generally assumed that the nature of these arrangements is critical for women's employment opportunities. Existing explanations of how familial responsibilities get translated into women's labor market positions tend to emphasize temporal patterns and constraints. For example, in explaining occupational segregation, some theorists fasten on to the fact that childbearing and child care take some women out of the labor force for periods of time. Women's jobs are also considered to be more "flexible" in terms of schedules and therefore attractive to women who have primary responsibility for child care. Considerable attention also has been given to the fact that child care responsibilities dictate part-time employment for some women. We explore several of these links between familial responsibilities and labor market position in the Worcester context. Our basic argument is that in some cases the focus on the temporal characteristics of household arrangements and women's employment is misleading and that it is the spatiality of the practices that is critical. At the very least, family strategies to manage paid employment and child care must be seen as spatial, and not simply temporal, ones.

Mothering and breaks from paid employment

The effects that extended breaks from paid employment have on women's position in the labor force deserve careful attention. They are such a common experience for so many women. In our sample of 492 women, 352 (72 percent) had had children at some point, and 267 (76 percent) of the women with children had taken (during the time period recorded in the job history) or were currently taking a childbearing/rearing break. The breaks averaged 6.9 years although the duration differed markedly by women's ages, indicating that household strategies are changing and that it is increasingly less common for women to leave the labor force for extended periods of time. The average break for women under 40 was less than five years, compared to almost ten years for women between 40 and 49 and more than 15 years for women 50 or older (see also O'Connell 1990; Desai and Waite 1991).

It is important to recognize that not all women are equally likely to take extended child-related breaks from the labor force. The propensity to do so clearly varies by a woman's own educational and occupational standing. Women who have taken breaks for childbearing/rearing tend to have, on average, less education and to have begun their careers with lower-status jobs than women who did not take such breaks.[13] Despite the clear association between a woman's own occupational standing and propensity to take an employment break, there is no relationship between husband's occupational standing,[14] and a woman's likelihood of leaving the labor force for a period of time to care for children. We do find, however, that maternity breaks are more common among middle-income households (those with incomes between $30,000 and $35,000) than among lower-income (below $30,000) and higher-income (above $35,000) households.[15] These patterns suggest one way that the experience of motherhood in combination with paid employment differs among women: women with less education and lower-status jobs and in middle-income households spend more time at home as full-time mothers.

Women in lower-status jobs were likely to take extended breaks from the labor force because, in the absence, prior to 1993, of any form of job-protected maternity leave, they were forced to quit their jobs at the time of childbirth. The benefit of unpaid maternity leave has been available to a much greater percentage of women in skilled manual or professional/managerial occupations (about 71 percent of the Worcester women in these occupations had this benefit) than it has been to women in unskilled occupations or skilled nonmanual jobs (44 percent and 58 percent of whom, respectively, had access to unpaid maternity leave) (p = .004). Maternity leave benefits are even more closely tied to personal income: 93 percent of women who earned $25,000 or more annually had this benefit, but only 34 percent of women who earned less than $15,000 did ($p$ = .00). (To put this into perspective, only 20 percent of the sample were in the former, high-income group, whereas roughly half of the sample were in the latter.) Without the benefit of maternity leave, many lower-income women have no doubt been forced to take a "break." If this break has long-term repercussions, the very pattern of breaks that we document reproduces existing inequalities among women, as well as between some women and men.

The long-term effects of childrearing breaks are poorly understood. Nevertheless, the strategy that involves women leaving the labor force for childbearing and childrearing enters into a number of different theoretical rationales for women's position in the labor market. It is, for example, the centerpoint of human capital theorists' explanations of (and rationalizations for) sex-based occupational segregation. Human capital theorists (e.g., Mincer and Polachek 1978; Mincer and Ofek 1982; Polachek 1981) see such interruptions in a person's work history as a major reason for women choosing jobs in female-dominated occupations, which supposedly penalize women less than do other types of work for intermittent or discontinuous labor force participation. Labor segmentation theorists also tie women's location in the secondary labor market to "interrupted" work histories (Jordan et al. 1992).

The balance of current empirical research tends not to support the theoretical claims of human capital theorists; women who take extended breaks from paid work are no more likely to have concentrated their work experiences in female-dominated occupations (England 1982; Corcoran et al. 1983).[16] A number of other effects of temporary or extended periods out of the labor force have, however, been noted. (For each of these effects there is also contradictory evidence, possibly reflecting differences in geographical context, data sources, and operationalization of concepts.) Several researchers have documented the association between breaks and subsequently dampened wages (Joshi 1984; Jacobsen and Levin 1992); in a similar vein, others have found a tendency for breaks from the labor force to be followed by downward occupational mobility (Dex 1987; Perry 1988), although this latter trend very much depends on a woman's occupational status before leaving the labor force (Dex 1987; Brannen 1989; Perry 1988; McRae 1991).

As important and practically useful this research is – for example, women's groups in the United States have used this type of analysis to lobby for federal Family and Medical Leave legislation to enable women to maintain their employment connection through childbirth and the first months of child care (Spalter-Roth and Hartmann 1991) – much of it conceives of women's lives in temporal but decidedly aspatial terms. When one examines this household strategy paying close attention to spatial relations, some interesting patterns appear.

From one perspective, that of a metropolitanwide map, the geography of this strategy is not particularly interesting. This is because, despite the association between household income and propensity to take breaks, women in households throughout the region take child-related breaks. (Remember that even in lower-income and higher-income households half of the women took child-related breaks.)

When one looks at the geography of individual women's lives, however, a very interesting pattern emerges. We find that for many women a child-related break is the critical juncture after which the spatial reach of their labor market is much restricted. The job that many women find after a break is much closer to home; the average daily commute to work (one way) before a family-related break was 19 minutes, significantly longer than the average 14.3 minute commute to the jobs women held after their breaks (Table 5.1). These mean distances understate the case for a large number of women: roughly half (44 percent) of the women in our sample took a job after a child-related break that reduced their commute by at least ten minutes.[17] Moreover,

Table 5.1 Travel-to-work times before and after a child-related break from the labor force

	Before	*After*	*Difference significant at 95%*
All women	19.0 (14.6) $n = 162$	14.3 (12.7) $n = 164$	Yes
Job after break full time	20.3 (13.2) $n = 64$	15.7 (11.3) $n = 69$	Yes
Jobs before and after break part time	14.5 (7.6) $n = 35$	16.7 (7.9) $n = 32$	No
Job before break full time/ after break part time	19.1 (15.6) $n = 61$	12.4 (16.4) $n = 58$	Yes
Job after break female-dominated	18.1 (12.0) $n = 109$	12.3 (7.6) $n = 109$	Yes
Job after break *not* female-dominated	19.4 (15.7) $n = 53$	15.8 (13.4) $n = 55$	No
Job after break professional/ managerial	15.7 (12.6) $n = 30$	13.5 (7.5) $n = 26$	No
Job after break skilled	17.7 (11.5) $n = 75$	13.5 (12.1) $n = 76$	Yes
Job after break unskilled	21.0 (15.5) $n = 57$	16.3 (14.6) $n = 62$	Yes

Source: Personal interviews, Worcester, MSA, 1987
Note: Women's average commute times in minutes; numbers in parentheses are standard deviations

the move to shorten the journey to work after a break is not simply an artifact of a shift into part-time work; women working full time after a break are also working closer to home than they did before the break (Table 5.1).

This spatial strategy of time management has quite dramatic effects on subsequent occupational mobility. Women who shorten their commutes after a childbearing/rearing break are more likely to experience downward occupational mobility. One in five women who shortened their commute by at least ten minutes experienced downward occupational mobility, compared to one in twenty women who did not shorten their commutes by this much time.[18]

We find, therefore, that women in lower occupational grades are more likely to take a break from the labor force after they bear a child, that such women search for jobs closer to home following a break, and that this is often associated with downward occupational mobility. Our argument is that, although theorists have focused on the temporal aspect of this strategy – the discontinuity in labor force participation – the spatial dynamic is at least as important. Women who leave the labor force for a period of time often search for employment in a much more circumscribed spatial range when they return to paid employment. Given the likely restriction in opportunities that accompanies this reduced range, is it surprising that a stagnation in occupational mobility is the typical result? Our findings supplement and qualify the research that has suggested that breaks from the labor force are associated with downward occupational mobility: this is especially true of women whose breaks coincide with a reevaluation of the distances that they are willing or able to travel to paid employment. Downward occupational mobility is a more likely scenario for those women who search for employment closer to home.

Are female-dominated jobs more flexible?

In addition to reducing the commute, choosing jobs with flexible hours is another reasonable strategy for juggling the demands of family and paid work. If jobs in female-dominated occupations do in fact offer more flexibility, then this explicit accommodation to women's double burden could help explain occupational segregation. It is clear that women do place a higher value than men do on having jobs with flexible hours: 26 percent of women in our sample (but only 16 percent of men) said that flexible hours was among the three most important attributes they looked for in a job. Similarly, in the recent National Study of the Changing Workforce (Galinsky *et al.* 1993), 44 percent of female, but only 26 percent of male, parents of children under the age of 13 said they would be willing to sacrifice job advancement for more flexible work arrangements involving flextime.

But this gender difference in attitudes toward job flexibility does not translate into occupational segregation. Women in all job types in our study were equally likely to say having flexible job hours was important or very important to them in evaluating job options (about three quarters of all women said this). Nor were women in female-dominated jobs more likely than other women to say that flexible job hours characterized their present job (about two-thirds of all women said they did). Indeed, the converse seems to be the case. When we explicitly asked about people's work schedules ("Do you have to be at work at a particular time or do you have some flexibility in your hours?"), a higher proportion of women in female-dominated occupations (38 percent) than women in male-dominated (21 percent) or gender-integrated (27 percent) ones said they had little or no flexibility in their hours of work.

This finding (that jobs in female-dominated occupations actually offer less flexibility) concurs with the results of other studies that have considered this question. Glass (1990) found that as percent female in an occupation increases, jobs have less flexibility, fewer unsupervised breaks, and are generally harder to perform. Similarly, in a study in Orange County, California that focused on whether or not employees could choose their own schedules, Giuliano et al. (1990) discovered that women are more likely than men to be in occupations that require fixed work schedules. Moreover, women are also more likely than men to have fixed schedules even when they are working in occupations where flexible schedules are generally more prevalent. Variations in the daily flexibility offered by different types of work cannot, therefore, help to explain occupational segregation. As we saw in Chapter 3, however, working part time – another strategy for balancing paid and mothering work and for easing the difficulties that rigid job schedules can pose – is strongly related to working in a female-dominated occupation.

Part-time employment

The strategy of part-time employment is integrally bound up with leaving the labor force for childbearing/childrearing: only a minority (26.1 percent) of women in Worcester worked part time immediately before a child-related break from the labor force; but immediately following such a break, 61.5 percent worked part time. Two jobs after a break, 52.4 percent of employed women were still working part time.

Despite the prevalence of part-time work among all mothers, those who work part time do tend to have particular household characteristics. First, this strategy is used more frequently by higher-status households.[19] Women in these households may work part time because their partner's wages allow it. A second characteristic of married women who work outside of the home on a part-time basis is that they have heavier domestic workloads than those who work in paid employment full time. As a rough way of summarizing each woman's domestic workload, we tabulated an index of household responsibility. A woman received a point for each of the following: the presence of children under 6 years of age; doing all of the housecleaning; having sole responsibility for cooking the meals; and doing most of the cleaning up after meals. Women with part-time employment tend to have much higher values on this index than those who are employed full time. Two-thirds of part-time workers had the highest possible score; in comparison, only one of three women working full time did all of these tasks ($p = .00$).

Perhaps because of heavy domestic workloads, part-time workers are (as we noted in the previous chapter), particularly likely to find employment close to home: the average home-to-employment commute is 13.6 minutes for women who work part time, compared to 16.8 minutes for women employed full time ($p = .02$). Women who work part time outside of the home in female-dominated occupations have the shortest commutes of all employed women. Their home-to-employment range is extremely circumscribed, with an average travel time of only 11.9 minutes (one way). It also appears that it is women who choose a part-time schedule as a strategy to combine paid with mothering work (i.e., those who move into part-time work only after a child-related break) who move within particularly constrained spatial orbits (Table 5.1). Finally, although almost all women working part time wanted a job close to home,[20] it does appear that those in female-dominated and gender-integrated jobs have been most likely to choose a particular job because of this job attribute. Most women working part-time hours in female-dominated or gender-integrated occupations said that "close to home" described their jobs well or very well (84.2 percent and 87.7 percent, respectively), whereas a minority (42.9 percent) of those in part-time, male-dominated jobs thought that this description was accurate.

What are the effects of electing to work in paid employment on a part-time basis? In evaluating this, our focus is one-sided, insofar as we assess only the effects on paid employment, and say nothing about levels of stress or sense of balance or personal fulfillment. Certainly very few women (only 13 percent) who presently worked part time wanted to move to full-time employment.[21] On the other hand, nearly one out of every three (30.3 percent) women with children at home who worked full time said that they would prefer to work part time. Contrary to the implications of others (e.g., Christopherson 1989) that women hold part-time paid jobs because they cannot find full-time ones, almost all of the part-time employed women in Worcester preferred part-time over full-time employment. (Pinch and Storey 1992 report a similar pattern in Southampton, U.K.)

Despite the desirability of a part-time schedule as a means of combining paid and parenting work, in practice long-term part-time employment has several negative consequences, most of which have been thoroughly documented (Beechey and Perkins 1987; Kishler Bennett and Alexander 1987; White 1983; Duffy *et al.* 1989; Perry 1988, 1990). Perry (1988) argues that returning to a part-time job has a significant negative effect on women's later occupational mobility. We have found that it is not simply the fact of returning to a part-time job after a break that is significant in itself. If we consider only those women who were working full time at the time of the interview, we find no significant wage or status effects of having returned to work part time after a break. That is, the wages and occupational status of women who had at first worked part time after a break and then moved to full-time work were no different from women who had always worked full time since a childbearing/rearing break. What does have a significant impact on wages and occupational status is the decision to *continue* to work part time – and this is the typical pattern: 67.7 percent of the women who re-entered the labor force into a part-time job after a child-related break were still working part time when we interviewed them in 1987.

Our analysis provides some support for the argument that the decision to work part time drives women into a ghettoized segment of the labor market, that of female-dominated jobs. This pattern is quite difficult to detect because of the strong overarching relationship between working part time and working in a female-dominated occupation.[22] The proportion of women working part time in female-dominated, gender-integrated, and male-dominated jobs, immediately before their last child-related break is 40 percent, 23 percent, and 17 percent, respectively. For the jobs that women return to (part time) immediately after such a break, these proportions jump to 64 percent for women in female-dominated, 51 percent in gender-integrated, and 38 percent in male-dominated occupations. Clearly, many women in all types of jobs choose to work part time as a way of combining mothering and paid employment, but the proportion is highest in female-dominated occupations.

Noting this tendency for most part-time workers to have jobs in female-dominated occupations, we can nevertheless detect a larger shift of workers from male-dominated and gender-integrated occupations into female-dominated ones among women who return to work part time rather than full time after a childbearing/rearing break. For example, for women employed in male-dominated occupations before a break, only 14.3 percent moved into a female-dominated job if they returned to full-time work, but fully 33.3 percent of those who returned part time shifted into female-dominated work. Considering those previously employed in gender-integrated occupations, 42.9 percent of those returning to full-time work entered a female-dominated occupation, compared to 63.6 percent of those who moved to part-time hours. The decision to work part time is, therefore, frequently associated with a shift into a female-dominated occupation.

These patterns may reflect the fact that part-time schedules are more readily available in female-dominated occupations. Within our 1987 Worcester sample, 33 different female-dominated occupations offered part-time hours. (This represents fully 72 percent of the female-dominated occupations represented in our sample.) In contrast, only 24 (35 percent) of the gender-integrated and seven (10 percent) of the male-dominated occupations had anyone in them working part time.

The drift of part-time workers into female-dominated occupations is worrisome because wage levels are particularly depressed for part-time workers in these occupations. Women who work in female-dominated occupations as part-time workers earned on average $6.58 an hour in 1987, while those who worked full time (and also had taken child-related breaks) earned $8.69 an hour. Women part-time workers in male-dominated occupations (only five in total!) actually earned more than full-time workers ($13.84 an hour as opposed to $10.79) and there was no wage difference between part-time and full-time workers in gender-integrated jobs ($9.57 as opposed to $9.61 an hour). Occupational mobility is also depressed for part-time workers.[23] These findings underline a common feminist position: there is an urgent need to restructure part-time work, such that a person who chooses to or must work part time is not thereby ghettoized in a low-paid segment of the labor market, with reduced chances for occupational mobility over her lifetime.

Our argument about the spatiality of part time work is that, given heavy domestic responsibilities, many married women with younger children elect to work in paid employment part time and close to home. Because part-time employment is more readily available in female-dominated occupations, those who seek part-time work and must find work close to home are especially likely to work in female-dominated occupations. As we have described in the previous chapter, an abundance of such jobs close to home increases the likelihood that a part-time employed mother will work in gender-typical occupations. Those who do, receive the lowest wages of all women workers, and are unlikely to experience upward occupational mobility.

Sequential scheduling

Many Worcester households manage parenting and paid employment by using another "temporal" strategy, that of arranging the parents' paid employment sequentially so that one adult can always be home to care for children (see also Lamphere 1987; D. Rose 1993 for discussions of this strategy). We have been struck by how many households in Worcester use this strategy. Sixty-six couples in our sample (representing 29 percent of all dual-earner households with children under 13 years of age) arranged their work in this manner.[24]

In three-quarters of the households that arrange their paid employment sequentially, it is done for the purpose of child care so that the family does not have to rely on an outside agent for this task. Sequential scheduling reflects a distrust of nonfamilial child care, as much as the expense and inadequacy of child care options in Worcester. The emotional and physical costs of enacting this family ideal are evident throughout the interviews, as, for example, in the case where husband and wife worked at the same establishment, located 40 minutes away from home. She worked an 8 a.m. to 4:30 p.m. shift; he from 2 p.m. to 10 p.m. He then went to a second job, where he worked until 2:30 a.m. "My wife works mornings. I work nights, so someone's home with the kids all of the time." They had maintained this schedule for 11 years and continued to do so for the sake of the children, despite the fact that their children were now aged 14 and 17. When questioned about his feelings about his wife's employment, the husband said, "I don't mind. She loves it. As long as it doesn't interfere with the kids. She could get a job on second or third shift, but I said no. She had to get a job in the mornings." His wife reflected, "We have to work our living schedules around our work schedules. We never worked days together because one of us had to be with the kids." Serial scheduling is, therefore, a household strategy that is adopted very self-consciously as a means of reconciling domestic and paid work.

The use of this sequential scheduling strategy has implications for timing as well as the type of paid employment chosen by different household members. It is significant that in the majority of cases it is the woman who takes the less optimal or less "conventional" time slot. In Worcester it is typically the woman who works the second or third shift; this was the case for 66 percent of the households whose members schedule their work sequentially. More than one-third of the women in households using this strategy spontaneously mentioned the effect that this schedule had on the type of job they could get, mentioning a compromise in the type of work or an inability to advance into management or administration. One woman bartender said: "I can't be a secretary and work 9 to 5. My kids come first so I have to work around their schedules." Another, who does keypunching from 5 to 10:30 at night for a local bank, commented that: "I love doing office work, running an office. Nothing like that is available in the evenings. You have to sidestep . . . you have to put the talents that you do have to work in the evening. Keypunching is not exactly my top choice." Another woman tells that: "When my youngest child was three, I had to go back to work for financial reasons. I wanted an 11 p.m. to 7 a.m. shift so that I could be with him [her son] when he was awake. So I looked for a job with those hours. I knew that I could just go and work there as a nurse's assistant without training and that I could get evening hours. I love doing it. I was just trying it out at first, but I love it. At the time I thought that being a nurse's assistant was about the only thing that I could do with those hours." As one further example, a woman who works from 3 p.m. to 11:30 p.m. as a medical technologist reported that she had to turn down a promotion because the hours of the higher-level job involved day work.

These data suggest that the gender characteristics of available second and third shifts have implications for the type of paid employment women do as well as women's participation in workplace organizations and career advancement. Indeed, 65 percent of women in households in which wife and husband arrange paid employment sequentially work in female-dominated jobs; this compares to 51 percent of women in households that do not use this sequential scheduling strategy ($p < .10$) The greater propensity to work in female-dominated jobs may well reflect the fact that women in households that use sequential scheduling are also more likely than are other women to work part time ($p < .05$).

Given our discussion of the geography of part-time employment, it is not surprising that sequential scheduling is not only a temporal but also a spatial strategy. Almost one-third of the women interviewed in the households using these strategies spontaneously mentioned that their schedules led them to work close to home. In some cases this is because they were reluctant to travel long distances at night. In others it is because the household work schedules are very closely timed, precluding lengthy work trips. For example, in one household the woman works as an office clerk from 8 a.m. until 1 p.m. in the afternoon. Her husband watches their three young children while she works. This woman had to select a job close to home (her current job is a 5–10 minute drive from home) because her husband has a 40-minute journey to work and his shift as forklift operator in a warehouse begins at 2 p.m. Sequential work trips place special temporal and hence spatial constraints on the journeys to employment of each parent, but it is most often the woman's commute that is the more constrained of the two.

It is readily arguable that we see patriarchal relations played out through these serial scheduling arrangements. Women tend to arrange their paid employment around the schedules of their husbands and children. In the words of one female interviewee: "I had to arrange my own job around everyone else's schedule." Women's subordination to the schedules of other members in the household clearly has some implications for their subordination within the paid labor force, as when preferences about the type of paid employment are totally subordinated to the suitability of hours. In many cases, career ambitions are set aside, at least for a time. The spatial limits of many individuals' labor markets shrink to a narrow circumference surrounding the home.

As one striking example of this tendency for women to accommodate to the schedules of other members of the household, one woman we interviewed was currently unemployed because her husband had elected to change his schedule from evenings to days. Presently aged 30, she had worked for five employers in the last seven years. She had started her work history at a large computer firm: "I would have stayed there and had a career. But if you want to raise your own child, you have to work part time and it makes it rather difficult." (Her first employer did not offer part-time work.) She then returned to a job that involved the sales and cleaning of carpets because "it had the hours that were important to me. [I was looking for] whatever fitted around my husband's schedule." After she had worked there for nine months, the company went out of business. Next she tried her hand at clerical work, filing and typing, because "it was in the papers and had the hours I wanted." Nine months later she was laid off. She next took a job as an assembler in an automobile brake factory: "It had 'mothers' hours [9–1] . . . they advertised it. If school was called off, they didn't expect us to come in." She left that job after a year and a half because of low wages and because "the piece work pissed me off." She then took a job as a deli worker, slicing and selling meats: "They had the right hours." After ten months at this job she was drawn back to the previous job as an assembler "because they offered me more money to come back and the hours were right." Having achieved a victory of sorts over the question of wages, she was forced to resign after two months in that job because her husband changed the hours of his job from nights to days. She was once again searching for a job: "I'm looking for a job with hours I can work . . . I'll take anything . . . probably factory or clerical." As an added constraint, she was restricting her area of job search to a 15 minute radius around her home because she wanted to be close to home for the children "to be able to get to them if something comes up." It is clear that this woman's work priorities have been tailored to her family's needs and schedules.

As with some of the other strategies that we have reviewed, such as the intergenerational transfer of property, sequential scheduling is far more common among working-class households: few wives of professionals and managers are leaving their homes in the evening for paid employment. Sequential scheduling is most common among families in which the male head of household is employed in a manual job. Considering only the occupations of the fathers in the households that use this strategy: 32 percent have jobs that can be classified as nonskilled manual, 9 percent as nonskilled nonmanual, 29 percent as skilled manual, 23 percent as skilled nonmanual, and only 5 percent as managerial/professional.

By calling attention to these class differences we do not mean to imply that patriarchal relations are not enacted in heterosexual households of all classes. We have seen that different strategies are employed in different households: the intergenerational transfer of property and sequential scheduling strategies are clearly working-class ones, in terms of the husband's occupational standing. Part-time work, on the other hand, is more prevalent among wives of men employed in higher occupational grades, most probably because the higher male income makes such a strategy financially feasible. The tendencies to take extended breaks from the labor force to care for children, to search for a job from an established residence, and to work part time are directly related to a woman's own occupational standing and are much less prevalent among women already employed in professional and managerial occupations. We have argued that these various family strategies, more common among women in nonprofessional, often female-dominated occupations and in working-class households, effectively lock them within these employment niches and that geography enters into the process that makes this so. The various family strategies reviewed here almost always lead the women in the households to search for jobs very close to home, usually in a labor market that bears little relation to their job needs or desires.

The last strategy reviewed does, however, suggest the complexity of patriarchal relations within households and the necessity of being attentive to variations and change. The men in households that organize employment sequentially are taking sole responsibility for child care for a portion of each day. In some (admittedly only nine) cases it is the male who takes the second or third shift so that he can be with his children during the day. In the words of one father who had voluntarily worked a night shift (11 p.m. to 7 a.m.) as a supervisor at a supermarket for the past ten years: "I don't trust anyone else to raise them . . . not even a relative . . . especially for the moral training." In other words, men are contributing a considerable amount of domestic labor (and, in some cases, accommodating their paid employment to the family's needs) so that women can take employment outside of the home. We turn now to explore with more care household circumstances that vary from the norm.

DIFFERENT FAMILIES, SIMILAR STORIES

Feminist theory has been fundamentally rethought over the last decade as more and more attention has been focused on the relations of domination written into a feminist theory that highlights gender to the exclusion of other differences such as race and sexual orientation; it tends to marginalize the experiences of some women and to universalize those of white, middle-class heterosexual women. Numerous feminists have argued that, aside from bearing the markings of race, class, and sexual privilege, the tendency to homogenize women's experiences places a feminist theory that ignores differences squarely within masculinist discourse: in Gayatri Spivak's works, "The question of woman in general is their question, not ours" (1983: 184). Gillian Rose (1993) has developed this argument within geography, noting the tendency of feminist geographers to universalize the experiences of mothers with young children in our understandings of the gendering of geographies.

Certainly women's domestic responsibilities and one family form (nuclear family, heterosexual couple with children) have figured prominently in explanations and rationalizations of women's subordinate labor market positions, both feminist and nonfeminist. Among the latter, human capital theorists, for example, search for the rationality of women's low wages and occupational standing in their movement into and out of the labor force during their childbearing years. Some male trade unionists have legitimized their own claims to a "family wage" and women's low wages through their selective portrayal of women as mothers of small children and dependent wives (who, if in paid employment, work only for "pin money") (Cunnison 1987). Anticipating that women will leave employment for childbearing/rearing, employers have indiscriminately slotted them into certain low-wage, dead-end jobs (Peterson 1989); similarly, employers, male workers, and unions have justified the exclusion of all women from certain job categories on the grounds that the job poses a danger for fetuses. For example, as recently as March 1991, the U.S. Supreme Court ruled in United Auto Workers v. Johnson Controls that employers cannot bar women from certain jobs on the grounds that such jobs involve exposure to substances dangerous to potential offspring.

In the previous sections we have tried to illustrate how marriage and children have a nontrivial impact on many women's (and men's) lives, one that reverberates throughout them. A description that highlights a single period in one type of household is, however, partial. Its partiality now seems more obvious and increasingly problematic. This is so for a number of empirical, theoretical, and political (strategic) reasons.

First, childbearing absorbs fewer and fewer years of the average woman's life, and there is now considerably more flexibility in the timing of these years than in the past (Davis 1988: 72; Phillips 1987: 60; Stacey 1990: 9). Second, "the nuclear family" and women's dependency on the male wage earner seem to describe the experiences of fewer and fewer people. Third, family composition is now extremely varied. For example, by 1988 one out of every four U.S. children lived with a single parent (Stacey 1990: 15). This statistic only begins to signal the fluidity and complexity of U.S. families, which also take blended, intergenerational, and communal forms. Stacey coined the term "post-modern family" to convey the sense of variable family strategies. "An ideological concept that imposes mythical homogeneity on the diverse means by which people organize their intimate relationships, 'the family' distorts and devalues this rich variety of kinship stories" (Stacey 1990: 269).

There are also theoretical and strategic reasons why feminist scholars should choose not to impose this homogeneity on our understanding of women's labor market experiences. One flows from a careful reading of how employers and male employees have attempted to exclude women from particular jobs by overgeneralizing their difference, as defined by childbearing capabilities. As Fincher (1993) has demonstrated for the Australian context, the state, through its policies, also tends to envisage all women as mothers of dependent children, often making labor force participation more difficult for women who do not fit this mold (such as those with elder care responsibilities). Another reason for analyzing household diversity is the recognition that feminist scholars also have tended to overgeneralize the links between sex and a specific set of household relationships in explanations of occupational segregation. This has led to a possibly essentialist link between gender and a particular set of occupations (Siltanen 1986). Siltanen has attempted to redirect feminist inquiry into inequality in the occupational structure by demonstrating that sex per se is less important as a condition for entry into a female-dominated occupation than is a particular set of domestic responsibilities (which may be held by men as well, though typically are not). In other words, she is suggesting the need to specify more closely the particular domestic experiences that lead one to take employment in what she calls "component wage" (as opposed to "full wage") jobs, and to disentangle these experiences theoretically from the categories woman and man. Siltanen's analysis not only sharpens an explanation of occupational segregation, it provides a framework for understanding differences among women, as well as common ground between some women and men.

Given our commitment to examining differences among women, we must nevertheless be explicit about the limited extent to which we are able to do this – reflecting limitations wrought by the boundaries of our conceptualization at the time of collecting our survey data, as well as by the characteristics of Worcester itself. First, as we have already noted, we do not have the resources to pursue an interest in variations by race. Given our decision to study a representative sample of Worcester households, we have very small numbers of Latino, African-American,

141

and Southeast Asian households in our sample. Second, in the interview process, we were relatively inattentive to questions of sexual orientation and how this might affect labor market experiences. We are relatively silent, therefore, on the questions of race and sexual orientation, undoubtedly the two most significant axes of difference discussed in contemporary feminist theory. Making a virtue out of necessity perhaps, our data do allow us to open up to discussion other differences that are relatively neglected in recent feminist theory: differences of class, age, arrangements of domestic work, and family form. We begin by assessing variations within a seemingly homogeneous category: dual-headed heterosexual nuclear families.

Different patterns within "conventional" families

In previous sections of this chapter we have traced the many ways that patriarchal relations within households and women's primary responsibility for domestic work shape and constrain women's employment options. Our focus has been on the prevailing pattern, which is rooted in patriarchal relations. In highlighting the predominant pattern, however, we have suppressed patterns that deviate from the "norm." Our focus now shifts to these other, less prevalent patterns, to differences among women and among men in heterosexual nuclear families. We by no means wish to qualify our argument that patriarchal relations are alive and well in Worcester and that they are enacted through spatial relations. At the same time, it seems strategically important to recognize the variety of experiences that exist within nuclear families. As Stacey (1990) notes, a static and overly simplified reading of the nuclear family downgrades the real impact that feminist political action has had on family life over the last twenty years or so. And as relatively rare as sightings of "the New Man" may be, it seems important to document them, to signal possibilities of social change within heterosexual nuclear families.

Heterosexual couples in Worcester choose many different ways of organizing domestic life and work. We attempted to capture these variations by asking about the division of domestic work within the household: who, for example, cooks the meals, washes the dishes, cleans the house, mows the lawn, runs errands, repairs the car? The division of household tasks is starkly divided along gender lines (Table 5.2): women carry out the bulk of the chores having to do with cooking, cleaning, shopping, and some aspects of child care, whereas men bear the brunt of yard work, and house and car repairs. In this regard the women and men of Worcester mirror a widespread pattern recently documented in the *National Study of the Changing Workforce*: women do the majority of daily household tasks even when they earn more than half the household

Table 5.2 Gender differences in household tasks

			Percentage of group saying "self always" or "self mainly" did task		
	Women (n = 268)	Men (n = 145)	Women in female-dominated occupations (n = 154)	Women in gender-integrated occupations (n = 96)	Women in male-dominated occupations (n = 18)
Cleaning the house	73	7	79	69	39
Cooking the meals	78	13	78	66	67
Cleaning up after meals	54	8	60	45	50
Shopping for food	68	18	68	69	61
Doing the laundry	77	10	80	74	72
Shopping for clothes	73	6	75	69	78
Paying bills	59	34	62	59	44
Staying home when child is sick	64	6	67	57	70
Chauffeuring the children	56	11	58	47	77
Pet care	43	18	50	35	21
Running errands	45	18	46	43	44
Mowing the lawn	6	67	8	6	0
Other yard work	13	55	10	12	22
House repairs	9	61	9	9	11
Car repairs	1	52	1	1	6

Source: Hanson and Pratt, 1990
Note: Responses are from employed individuals in couple-headed households. Respondents were given the following categories: you always; you mainly; partner mainly; both equally; someone else

income. In addition, younger men contribute no more to domestic work than do older ones (Galinsky *et al.* 1993). This underlines our point that an interest in exploring differences among women and men should not blind us to the continuing overwhelming predominance of patriarchal relations.

At the same time, different groups of women do have very different domestic workloads, depending on their and their husband's occupational status, signalling that responsibility for domestic work reflects power relations, within and beyond the household. Men with higher personal incomes are less likely than other men are to help with domestic work ($p = .01$). Having a wife who is in the paid labor force, however, markedly increases not only the likelihood that a man will take on some domestic work but also the number of these different chores he is involved in. Among the 92 married men in our sample were 72 who said that they assumed some responsibility (either sole responsibility or that equal to their partner) for one or more of the following domestic chores: cleaning, cooking, kitchen cleanup, and childcare.[25] For those with working wives, less than one quarter (24 percent) said their partner bore the entire responsibility for all household chores. This compares with 34 percent of men whose wives were not in the paid labor force ($p = .03$). And 44 percent of men with employed wives versus only 20 percent with wives out of the labor force took equal or primary responsibility for two or more chores. (Morris 1994 reports a similar pattern in the British context.)

Patterns of male participation in domestic work are tied, not only to wives' participation in the labor force, but to the type of occupation in which the woman is employed.[26] Table 5.3 shows that there is a real range of responsibility for domestic work among employed married women and that this is related to characteristics of their paid employment.[27] Women who are less likely to take sole responsibility for domestic tasks are more likely to work in gender-integrated or gender-atypical occupations. Men in households in which women are employed in male-dominated occupations are especially likely to contribute domestic labor. Almost one in four men in these households (21 percent) took sole or "main" responsibility for at least two of the four previously mentioned tasks. Only 8 percent of men married to women in female-dominated occupations and 6 percent of men married to women in gender-integrated occupations did this much domestic work ($p = .003$). Although the direction of causation between domestic workload and women's occupational type is difficult to disentangle (i.e., does a job in a male-dominated occupation precede or follow a more equitable distribution of tasks within the household?), this finding brings us back to Siltanen's point: it is not gender or even marital status, per se, that is

144

Table 5.3 Household responsibilites of married women and partners

	All employed women in			Men with wives in		
Index score	Female-dominated occupations	Gender-integrated occupations	Male-dominated occupations	Female-dominated occupations	Gender-integrated occupations	Male-dominated occupations
0	3.9%	6.2%	11.1%	45.7%	41.3%	52.2
1	15.7	27.8	33.3	46.3	52.9	26.1
2	20.3	29.9	16.7	7.4	2.9	13.0
3	43.1	25.8	33.3	0.6	2.9	8.7
4	17.0	10.3	5.6	0.0	0.0	0.0

$\chi^2 = 17.89$; $d.f. = 8$; $p = .02$ $\chi^2 = 14.29$; $d.f. = 6$; $p = .02$

Source: Personal interviews, Worcester, MSA, 1987
Note: Index score ranges from 0–4. For women, individual gets a point for each of the following: has children under 6 years old: does all of the cleaning of the house; cooks all of the meals; cleans up after meals. For men, individual gets a point if he takes all or main responsibility for each of the following: cleaning the house; cooking meals; and cleaning up after meals. He also obtains a point if his partner or he says that he takes sole responsibility for childcare at some points in time or simply assists with the children

Table 5.4 Married women's evaluation of job attributes in general

| | Percentage saying attribute is important or very important | | |
	Mature married women	Other married women	p-value
Close to home	85	79	
Easy to get to	86	78	
Low transportation expense	62	56	
Proximity to child care/schools	58	72	
The job hours	72	86	.01
Having flexible hours	63	80	.01
Possibilities for advancement	61	72	.05
People you work with	91	87	
The type of work	94	93	
Good pay	88	92	
Good benefits	85	76	
Amount of independence	92	80	.02
Amount of prestige	47	41	
Physical work environment	70	67	
Fit with partner's work schedule	79	81	
Fit with child care/school	63	88	.01

Source: Pratt and Hanson, 1993
Note: The *p*-value is from a χ^2 test contrasting the two groups' evaluation of each attribute as either (1) very important or important; (2) neither important nor unimportant; and (3) unimportant or very important. Mature women were defined as those who were married, over 40 years of age, and had no children under 18 living at home

significant for understanding occupational segregation but, rather, a more specific set of domestic relationships.

The difference in domestic workloads for women in different occupations is most likely tied to the geography of paid employment, more specifically, to the fact that women in female-dominated occupations tend to find jobs so close to home. With fewer domestic responsibilities, women in other occupations also live different geographies of paid of employment, being much more likely (as noted in Chapter 4) to find paid employment farther from home. But the impress of domestic workload on labor market geography is felt not just by women; it is felt also by men. Men who participated in domestic labor worked significantly closer to home (their average work-trip travel time was 18.9 minutes, $s = 13.1$) than the twenty married men who said they were not at all involved in such chores (their average travel time was 26.4 minutes; $s = 17.3$; $p = .04$).[28]

The relationship between domestic workload, commuting time, and occupational type is, however, a complex one – even within conventional nuclear families. We find, for example, that the tug of home continues to be felt by middle-aged married women who are no longer actively

146

involved in childrearing. Most mature married women (whom we have defined to include those who are over 40 and who have no children under 18 at home) continued to value an employment location that is close to home (Table 5.4). They traveled no longer to work than did other married women (both averaged about 16 minutes), and they were no more likely to consider taking a job farther from home than were younger married women (for both groups the maximum commute they could envisage was about 27 minutes on average).

The sustained importance of proximity no doubt reflects the fact that some mature women continue to have heavy domestic workloads (Table 5.5, columns 2 and 3). As women move through the life course, they may face new kinds of family responsibilities not acknowledged in a theoretical perspective that fastens on the childbearing and early child-rearing years. The mature married women in our sample, while not typically responsible for young children, were more likely than were the other married women to have someone else who relied on them for care on a regular basis; about one-third (32 percent) of the mature married women, versus one-fifth of the other married women, said that someone, like a parent or a grandchild, depended on them for care on a regular basis. Many women in our sample found themselves, after their children were grown, altering their labor force participation to accommodate these new family demands.

The experience of one woman documents a life course of different but recurring responsibility for others. This woman, 50 years old when interviewed, left the labor force until her oldest child was 13 at which point she returned to paid employment in circumstances that allowed her to blend domestic and paid work. "I went to Wonder Market because it was part time, convenient, close to home, and the children were at school. My work hours were perfect for their school hours, and I could get home in an emergency. I worked because I wanted to get out of the house. It was the perfect job for me because it was close to home, and I got all of my accounting training on the job." She eventually took full-time employment as an accountant but at the time of the interview was planning to shift from full-time to part-time employment, to manage a new set of responsibilities: the care of her parents. "By working full time I am too tired to see them very much, and I feel guilty. I want to be able to spend more time with them and also have more time around the house." Many other mature married women had switched from full-time to part-time paid work, or had left the labor market entirely, to care for grandchildren or for spouse or parents.

The domestic workloads of mature women clearly affect their levels of labor force participation, with those who have lighter domestic work-loads much more likely to be employed full time (Table 5.5, columns 4 and 5). And, as is the case for all women, mature women with lighter

Table 5.5 Household responsibilities of married women

Household responsibility index	Mature married women	Other married women	Mature married women who work		Mature women's mean travel time to work in minutes	
			full time	part time	n	travel time
0	3 (3.2)[a]	10 (3.8)	3 (7.7)	0 (0)	3	17.7 (2.5)[b]
1	10 (10.6)	39 (14.7)	6 (15.4)	1 (4.3)	7	27.3 (22.7)
2	25 (25.5)	61 (22.9)	16 (41.0)	3 (13.0)	19	15.2 (9.4)
3	39 (41.5)	85 (32.0)	11 (28.2)	12 (52.2)	22	14.0 (9.4)
4	17 (18.1)	71 (26.7)	3 (7.7)	7 (30.4)	9	7.6 (9.0)
Totals	94 (100.0)	266 (100.0)	39 (100.0)	23 (100.0)	60	
	$\chi^2 = 5.21$; $d.f. = 4$; $p = .26$		$\chi^2 = 13.90$; $d.f. = 4$; $p = .007$		$F^2 = 3.04$; $d.f. = 4$; $p = .02$	

Source: Personal interviews, Worcester, MSA, 1987
Note: The household responsibility index was recalculated for Mature women, substituting care for persons other than children for the presence of a child under six. Mature women defined as over age 40, no children under age 18 at home
[a] Column percentage
[b] Standard deviation

domestic responsibilities are more likely to be employed in male-dominated jobs.[29] We can also see the effects of household responsibilities being played out in space (Table 5.5, columns 6 and 7): those with the heaviest domestic loads find employment very close to home, a mere eight minutes away, while those with the lightest loads travel the farthest.

In general, therefore, across different groups of women and men living within nuclear families, we find that those with the heaviest domestic workloads are most spatially constrained and, among women, are more likely to work part time, in female-dominated occupations. But the point must be made: there is enough variation in domestic workloads within nuclear families to allow us to detect these trends. We also call attention to the fact that a focus on domestic responsibilities and their effects on labor market experiences does not entail an exclusive focus on mothers with young children: caring for others – children, grandchildren, parents, grandparents, husbands, aunts – is central to the lives of many women and continues, for many women, throughout the life course.

Employment geographies of divorced, widowed, and never-married women

Explanations that tie sex-based occupational segregation to women's familial responsibilities and to patriarchal relations within households imply that some women – namely, women without children and male partners – are better placed to avoid their gendered destiny. We do find this to be the case. Never-married, divorced, separated, and widowed women (in comparison with married women) are far less likely to be employed in female-dominated occupations (Table 5.6).[30] This substantiates Siltanen's point that many of the occupations labeled as

Table 5.6 Percentage of employed women in each occupational type, by marital status

		Occupation type		
	n	*Female-dominated*	*Gender-integrated*	*Male-dominated*
Married or living with male partner	270	57.1	36.3	6.6
Divorced, separated, widowed	45	40.0	40.0	20.0
Never married	21	42.9	47.6	9.5

$\chi^2 = 11.34$; $d.f. = 4$; $p = .02$

Source: Pratt and Hanson, 1993

149

Table 5.7 Family circumstances of employed women

	Women with income >$35,000 per annum		Women in male-dominated occupations		Family circumstances of all employed women	
	n	%	n	%	n	%
Single	1	(9)	3	(10)	21	(6)
Currently divorced, separated	4	(36)	6	(21)	32	(10)
Married, no children	4	(36)	2	(7)	78	(23)
Widowed	0	(0)	4	(14)	13	(4)
Husband ill, disabled	0	(0)	2	(7)	n.a.	(—)
	9	(81)	17	(59)	144	(43)
Married with children						
Helping husband	0	(0)	4	(14)	n.a.	(—)
Professional status attained before marriage and children, career on hold	0	(0)	3	(10)		
Children grown	0	(0)	1	(3)	n.a.	(—)
Children school age or younger	1	(9)	4	(14)	n.a.	(—)
	2	(18)	12	(41)	191	(57)

Source: Pratt and Hanson, 1993
Note: n.a. indicates not available

female-dominated are better conceived of as "component wage" jobs, ones that comprise just a component of a household's income. The majority of women who head households on their own do not have jobs in these occupations. As further support for Siltanen's argument, the personal incomes of employed single, divorced, separated, and widowed women also tend to be higher than those of married women. For example, 48 percent of employed married women had personal annual incomes below $9,000 while "only" 33 percent of divorced, separated, and widowed women and 22 percent of single women had incomes in this range. Alternatively, almost one in ten divorced, widowed, and separated women had personal annual incomes of at least $35,000, compared to only one in fifty married women and one in twenty never-married women.[31] Income differences also reflect the fact that employed single, divorced, separated, and widowed women are much more likely to work full time in paid employment: this was the case for 89 percent of women in these circumstances, compared to 59 percent of employed married women. Women who head households also want different things from their paid employment than do most married women. They care more about opportunities for advancement and good benefit packages.[32]

The limitations imposed by marriage and children are strikingly obvious when we take a closer look at women who earn relatively high incomes or work in traditionally male occupations. Actually, there are very few of these women to study: only eleven (or roughly 3 percent) of the employed women in our sample had an annual income of $35,000 or more,[33] and only 29 were employed in male-dominated occupations. In Table 5.7 we show the family circumstances of women with high incomes and jobs in male-dominated occupations. Almost all women with high incomes have what can be termed "nontraditional" family circumstances: they are single, divorced, or married without children. Only one woman in our entire sample was juggling childrearing with a relatively well-paying job.

Women in male-dominated occupations are also unlikely to conform to the "norm" of marriage and children. Furthermore, the majority of women in male-dominated jobs who are married and have school-aged or younger children also have what might be called a "fragile" relationship to paid employment. A number of these women saw their employment primarily in terms of helping their husbands. One woman in this group, for example, was a wallpaper hanger who helped her husband in his wallpapering and painting business. This is how she described her labor force participation:

I can anticipate what he needs in a way that other people might not do. I take pride in his work, and I want to make it easier for him. So

Table 5.8 Travel-to-work times for women and men in different household types

Household type	Average time to paid employment	All employed women					Employed men	
		Occupational type			Care for others?		Care for others?	
		Female	Integrated	Male	Yes	No	Yes	No
Married/cohabit	15.5 (15.3) n = 258	12.8 (12.2) n = 145	17.5 (17.3) n = 92	27.3 (21.0) n = 18	13.7 (13.2) n = 60	15.7 (14.5) n = 186	17.8 (13.8) n = 18	20.9 (15.2) n = 115
Never married	14.4 (12.2) n = 22	10.2 (6.3) n = 9	17.1 (16.0) n = 10	18.3 (10.4) n = 3	11.6 (5.9) n = 5	16.3 (14.1) n = 15	11.3 (9.9) n = 6	27.1 (23.5) n = 19
Divorced, widowed, separated	15.5 (13.5) n = 44	12.5 (14.4) n = 17	16.1 (12.5) n = 17	19.7 (14.6) n = 9	9.7 (9.0) n = 8	16.8 (14.1) n = 36	n.a. – n = 0	5.6 (8.2) n = 10

Source: Personal interviews, Worcester, MSA, 1987
Note: Average travel time in minutes; numbers in parentheses refer to standard deviations for means in row above

basically [I do this because] it's the time I can work and I enjoy working with him. I give him speed.

To this woman, what was important about her job was that she is "helping him . . . serving him." Several other married women in male-dominated occupations with children of school age or younger mentioned that their careers were more or less "on hold" at the moment because they were presently working outside of the home on a part-time basis. In the words of an architect: "It's been my choice to have kids as a priority. My profession has suffered because of it. My classmates are already far along and I'm still working part time." What we have found, therefore, is that women who are exceptional in their paid employment also tend to have "exceptional" family circumstances.

What is perhaps troubling to our general thesis about the spatiality of occupational segregation, however, is that women in "nontraditional" households also tend to find paid employment close to home[34] (Table 5.8). There are two ways that these patterns seem to upset our thesis. First, the fact that most never-married women also find employment close to home suggests that familial responsibilities may have been overdrawn as a rationale for women's short home-to-employment commutes. Second, although single, separated, divorced, and widowed women are, on average, no more likely than are married women to travel long distances to paid employment, a larger proportion find jobs in male-dominated occupations. (And this is so despite the fact that married women's educational attainment does not differ from that of divorced, separated, and widowed women: see note 30.) This raises doubts about the importance of spatial constraint as an explanation for occupational segregation. (It is important to note, however, that, within each household type, women in female-dominated occupations spend the least time and women in male-dominated occupations spend the most time commuting (Table 5.8).)

With reference to the first point, regarding women who have never married, length of commute may reflect circumstances that differ substantially from those of other women. This is because never-married women were much more likely than other women to locate their residence *after* they had found a job: roughly one in three never-married women found their job first, compared to 5.9 percent of married and 4.4 percent of separated, divorced, and widowed women ($p = .000$). They likely choose to live near work. With reference to the second concern, the majority (68 percent) of single women live within the City of Worcester (versus 42 percent of married and 41 percent of separated, divorced, and widowed women: $p = .05$). This means that a greater variety of employment opportunities (including those in gender-atypical jobs) is more readily accessible to never-married women within a short commuting distance.

It is also important to remember that many never-married individuals are embedded within a network of social relations that may carry with them a set of obligations no less binding and time-consuming than those experienced by many married women. It is no less problematic to generalize across "single" households than it is to generalize about married women, and it is inaccurate to portray all single women (or men) as relatively free of domestic responsibilities.

One out of ten never-married women and men lived in an extended family, often living with and taking care of an older relative, usually a parent. Single women (and men) often have someone other than a child or partner to take care of on a regular basis: one out of four single women and men reported this responsibility.[35] A number of those interviewed who presently lived with an ageing parent or other relative mentioned the centripetal geographical tug exerted by these connections. From a 47-year-old woman who lived with her brother and sister-in-law in what was her parent's home, "Closeness and security of neighborhood, that's why I live here, and work follows from that." A 46-year-old woman who lived with her retired 82-year-old father and worked as an insurance clerk cited proximity to her parents as one of the major factors dictating her career moves throughout her lifetime. From a 49-year-old man who lives with his 83-year-old mother and 60-year-old brother (both of whom have Alzheimer's disease): "I would like to move but my mother doesn't want to."

The commitment to care for others in the extended family has an impact on labor market participation. Not one of the fourteen single, separated, divorced, or widowed women who reported caring for someone other than a child on a regular basis was employed in a male-dominated occupation.[36] And, returning once again to the role of spatial constraint, employed women who report the responsibility for caring for others in their extended family travel shorter distances to employment than those without this responsibility (Table 5.8). This is especially true for nonmarried women. The patterns for men are also fascinating, in particular, the tendency for single men who care for someone in their extended family to work close to home. This is especially interesting because the single men who claimed this commitment actually are responsible for it (they have no partner to perform the family "caring" function), and thus they may more closely approximate a stereotypically feminine caring role.[37]

Just as it is problematic to generalize across all married heterosexual couples, then, there is a messiness within the category of "never married" (as well as "separated, divorced, and widowed") and considerable variability in the amounts of domestic labor carried out by different

154

individuals in each household type. But, within this variability, we see some of the same patterns emerging: those with more domestic responsibilities tend to work closer to home and tend not to work in gender-atypical occupations.

CONCLUSIONS

We have tried to understand how patriarchal relations within households continue to limit women's employment chances and maintain sex-based occupational segregation, and we have tried to draw out the geography that underlies this. We have emphasized the importance of not overgeneralizing women's family circumstances, as when all women are treated – as they have been by some theorists, employers, and by government policy – as if they are the mothers of young children. In fact, the diversity of women's household circumstances is related to diversity in aspects of women's labor market participation, including its geography.

In concluding, we would like to underline the particularity of our narrative about the spatiality of household relations and employment. Our geographical tale is quite a simple one: those who fit paid employment around heavy schedules of domestic work, most often women, typically find employment close to home. There are a variety of ways of managing this (for example, continuous labor force participation or child-related breaks; simultaneous or sequential scheduling; part-time or full-time hours), but generally those who have the heaviest domestic responsibilities do work closest to home. It is unlikely that the residential locations of these women were chosen with either their employment or these spatial restrictions in mind. To some extent "trapped" in place, many women search for jobs from a narrowed range of possibilities.

This is a narrow reading of both the family and work. We recognize that child care and responsibility for extended families and friends are more than work; they can be means for and outcomes of rich and fulfilling lives. So too, there are many ways of valuing work that our measures of mobility, wages, and the gender-typing, class, and status of occupations scarcely begin to capture. A number of women and men told us about a restless reevaluation of the meaning of work throughout their lifetimes, as family and other circumstances changed. This might involve leaving paid employment to go to Cape Cod to paint for six months or to spend time with one's grandchildren or one's husband in his retirement. These kinds of reevaluation, of how to combine and value different kinds of work, pleasures, and commitments, are ones that we have not pursued, concerned as we are with women's position in the formal, waged labor market.

With some exceptions, women's positions in the formal labor market continue to be disadvantaged ones. We have shown how relations within families and households, lived through space, contribute to this fact despite the very real pleasures family life holds for many (not all) women.

6

EMPLOYER PRACTICES, LOCAL LABOR MARKETS, AND OCCUPATIONAL SEGREGATION

Many commentators have recognized that women's employment situations reflect more than their family circumstances; a number of complementary processes within labor markets and workplaces also structure women's employment opportunities. Our interviews with employers in Worcester offer some insight into the ways that employers shape how work becomes gendered. The interviews are, for example, replete with stereotypical observations about women's special characteristics that qualify or disqualify them for specific occupations and tasks. Women, we are told, are better suited for repetitive, precision tasks because they have dexterity and patience, and "men just can't hack sitting there." On the other hand, women were protected from some better-paid production jobs because the workplace was too hot or "not a feminine environment"; they were excluded from others because of their assumed inability to lift 50 or 70 lbs. (One woman who was employed as an extruder, a male-dominated occupation that required lifting 50 lb. bags of compounds, explained: "It's like picking up grocery bags.") We were also told that men are better as bill collectors because of their "stern outlook and voice on the phone. . . . People tend to procrastinate more with a woman than with a stern man." Racial stereotypes were no less prevalent. We were told by a manufacturer of refrigerated systems, for example, that African Americans were unsuitable employees because of their natural aversion to the cold.

Many of the employers use familial ideology to frame women workers; their stereotypes about women are tied to assumptions about the ways that family arrangements will affect their productivity as employees (Pratt and Hanson 1993). In general, employers know much more about the family circumstances of their female than of their male employees: one in ten employers could provide information only about women employees, and one employer told us that: "Marital status isn't as much a part of a man's identity." Only eleven out of 139 employers ventured an opinion about the workplace relevance of the marital status of male employees, and almost all of these employers preferred married men for stability and reliability. Twice as many employers made observations about the effects of marital status on women's productivity. They were divided in their preferences for married and single women; a slight majority favored single women, especially for professional and higher-status jobs. The one woman working as a sales representative in an advertising firm was described by her boss as "doing excellent. It's hard to find females who are single. The job doesn't mesh with school hours of kids. Our sales woman is mortgaged to the hilt. Nice car. Nice condo. She likes to work. She has a pleasant personality." (This employer stated a preference for hiring women: "Women in this business do a hell of a lot better than guys do. . . . Pretty women do better. It's selling the package. If it's a nice package you buy it." Despite these stated gender preferences, he has employed only one female salesperson (out of a total of 19), presumably because of his difficulty in finding women with no family responsibilities.)

Rather than expand on the multitude of ways that employer discriminatory attitudes and practices structure women's and men's work experiences – already well documented in the existing literature (e.g., Reskin and Roos 1990), we focus on the geography of employer strategies. We are particularly attentive to how employers' sensitivity to the availability of particular types of labor influences their locational decisions, creating distinctive labor markets and literally mapping labor market segmentation into place. Their hiring practices then reinforce the localization of distinctive labor markets, as well as gender- and class-based occupational segregation. At an even finer spatial scale, spatial segregation within establishments mirrors and reproduces occupational segregation.

In this chapter we focus on employer strategies, although we attempt to show how employers' strategies and labor practices are affected by the characteristics of employees and places. The decision to focus on employers first is somewhat arbitrary; in the next chapter our attention shifts to employees and the ways that their job searches complement the hiring practices of employers to create very local and distinctive labor markets. Underlying our general argument is an appreciation of the interrelatedness of social and economic life (for example, employers' stereotypes are saturated with familial ideology; workers' job searches are often an expression of their social networks of friends, neighbors, and kin) and the interdependencies between employers and employees. Clark (1983: 369) reminds us that "the labor market is a social construct, it is 'human' made! Consequently, it bears the imprint of the structure of power relations in society." These power relations are not likely to be unidirectional, and employers do not "hold" absolute power. Rather, these relations are worked out locally through a dynamic relationship between employers and employees, women and men, within homes and throughout workplaces.

We focus on employers and workplaces in the four labor markets within the Worcester metropolitan area that we described in Chapter 3. Bielby and Baron (1987), noting the variability of occupational segregation across establishments and organizations, have recognized the importance of attending to organizational practices within specific firms and the ways they contribute to the gendering of employment. By noting how these practices both structure and reflect very local and distinctive labor markets, we bring a geographical understanding to organizational practices and occupational segregation. We begin also to understand the way that labor market segmentation is always carried out at particular sites.

EMPLOYERS' LOCATIONAL STRATEGIES AND LOCAL LABOR MARKETS

For the majority of employers we interviewed, access to labor was important to their choice of current location.[1] Because more than two-thirds (67 percent) of our respondents said that they had not considered locating outside of the Worcester area, we know that the bulk of employers were surveying different locations within this metropolitan area, weighing the advantages and disadvantages of various sites. Most employers mention the importance of site costs and characteristics and access to transportation, but the majority in each area also mention the importance of access to labor – in particular, either skilled or inexpensive – to their choice of current location (see also Nelson 1986).

Storper and Walker (1983) have argued that segmented labor markets have a clear spatial expression. The fact that employers with very different labor requirements seek out different areas within the metropolis gives us a sense of how this labor market segmentation is mapped onto places. Employers in different areas vary both in the value that they place on access to skilled as opposed to inexpensive labor (Table 6.1) and in the particular skills that they want from skilled workers (Table 6.2). Few employers in Westborough desired access to inexpensive labor, but this was important to a good number of employers in the other areas. And while the majority of employers in both Westborough and Main South sought out labor with appropriate skills, they were clearly in search of different types of skilled labor (Table 6.2), with Westborough employers most often interested in accessibility to engineers, and Main South (as well as Upper Burncoat and Blackstone Valley) employers especially concerned about skilled production workers.

Although the labor requirements of employers in areas other than Westborough look quite similar in Tables 6.1 and 6.2, the workers available in each area are quite different; employers were keenly aware of this and indicated their knowledge and appreciation of characteristics

159

Table 6.1 Factors affecting firms' location decisions

	Main South (n = 39)	Upper Burncoat (n = 14)	Blackstone Valley (n = 30)	Westborough (n = 42)	p-value
Cost of land/building	76.3	76.9	83.3	71.4	n.s.
Physical appearance/ public image of location and facility	33.3	35.7	27.6	59.5	.03
Availability of an inexpensive labor pool	46.2	35.7	50.0	16.7	.00
Availability of labor pool with appropriate skills	53.8	57.1	34.5	66.7	.08
Proximity to public transportation	46.2	14.3	0.0	4.8	.00
Proximity to executives' home	13.2	7.1	43.3	42.9	.01
Site characteristics	56.4	35.7	60.0	59.5	n.s.
Proximity to transportation for delivery of goods	48.7	71.4	50.0	64.3	n.s.

Source: Personal interviews with Worcester area employers, 1989
Note: Respondents were asked to say how important, on a scale of 1 to 5, each factor was in the firm's choice of its present location. Numbers refer to percent of employers in area saying factor was "important" or "very important" in decision to take present location. The p-value is from a χ^2 test for differences among areas

specific to the workers in their area. A number of the Upper Burncoat employers had been attracted to the industrial park because of its proximity to the neighboring public housing project; they were aware of the indirect subsidies in the form of inexpensive housing and child care facilities. At an electronics firm we heard:

> Wages in the electronics industry were low, the work was light and repetitious. It required women who liked to do sewing, stitching and knitting. We need women with good eyesight who don't mind working with a microscope. . . . The property was reasonable and there were tax concessions. Available labor was a big consideration. . . . Great Brook Valley was an attraction; they had 600 units.

A plastics manufacturer tells a similar story: "I thought at the time [of selecting the site] that the housing project across the street would provide the ideal labor market, low-cost residences." Another manufacturer told

160

Table 6.2 Type of labor important to firm's location decision

	Main South	Upper Burncoat	Blackstone Valley	Westborough
Unskilled/routine production	52.0 (47.3)	53.3 (40.5)	62.5 (42.6)	28.2 (13.3)
Skilled production/ supervisors	64.0 (28.0)	60.0 (26.0)	66.7 (18.3)	18.8 (4.8)
Engineers	0 (0.7)	13.3 (3.0)	0 (2.6)	37.5 (18.1)
Clerical/office	4.0 (9.1)	6.7 (6.4)	0 (13.2)	12.5 (10.9)
Sales/marketing	0 (3.2)	0 (8.9)	0 (9.9)	15.6 (28.2)
Professional/ managerial	8.0 (11.7)	6.7 (15.2)	0 (13.4)	25.0 (24.7)

Source: Personal interviews with Worcester area employers, 1989
Note: Numbers refer to percentage of firms in this area that mentioned this type of labor as important; some employers mentioned more than one type; numbers in parentheses are the actual proportion of the area's labor force (in the interviewed manufacturing and producer services firms) in this occupation

us that: "The fact that Great Brook Valley provides child care is a boon to us because it allows people to work here. . . . We use the Great Brook Valley day care."

Employers in the Blackstone Valley understood their local labor force in terms of skill, gender, household assets, and the work culture of the area. An owner of a cleaning service in the Blackstone Valley explained:

We are looking for cheap labor, unskilled labor. There are certain advantages to unskilled labor; it's a simple job, it just takes a lot of common sense. An overskilled person would be detrimental. They don't last very long; you can mold an unskilled person to what you want. All of our labor force are women and all work part time. It's attractive to them because you can come in and work two or three hours and get home before the kids get back from school. I considered locating in Worcester; there's a better labor base there. But most of the people I could hire in Worcester don't have vehicles. You have to have a car to work this job; they get paid mileage.

From the owner of a chemical factory who had moved to the Blackstone Valley in the 1960s we heard:

The labor pool was magnificent, very bright older people. The textile mills had left. You had people who give a damn about their work. They had general factory intelligence. They were well trained, well disciplined, willing to work overtime. [They had] an excellent work ethic. They were very frugal people. They lived within their means.

(In these last two statements this employer may have been acknowledging the Valley's long-standing culture of low wages.)

Finally, employers attracted to Main South were well aware of the presence and immobility of various immigrant populations. For example, a Main South footwear manufacturer told us:

Four years ago the company, with headquarters in Fitchburg, decided to open a new branch in order to take advantage of the Spanish population in Main South. Because of the availability of this labor pool and the lack of unions, [we] decided to open a small operation here. The company used this as a pilot project to see if it is successful. The footwear industry being what it is, we require experienced labor. Because of hard times, two local shoe companies within two blocks of this location had closed down, and we took on some of their labor force. We knew there was an available experienced labor force here because these companies had closed down, and we actually got about half of our labor force from these companies: we were looking for immigrant labor, skilled in stitching and lasting.

We begin to see, then, a process through which fairly contained, distinctive labor markets and employment niches develop throughout the city. Employers are drawn to certain districts because of the availability of workers with particular skills, household assets, and desired ethnic, gender, and work-based cultural characteristics. Through this process we see labor market segmentation mapped onto places. Certainly, as Peck (1989) observes, there is also labor market segmentation within places, such that men and women, Latinos and whites, and managers, clerical workers, and production workers – all living within the same area, such as a city – often do not compete for the same jobs. But at a smaller geographic scale (i.e., within urban areas) and across particular occupations, there is a sharp spatial segregation that mirrors labor market segmentation. As shown in Table 6.2, the occupational composition of the labor force in the sampled firms varies by area. Hardly any sales and engineering jobs, and relatively few professional and managerial ones, are to be found in areas other than Westborough. Production jobs, on the other hand, are much more readily available in these other areas.

Further, the labor market segmentation of particular ethnic groups has a spatial expression, so much so that an employer located in Upper

162

Burncoat can remark casually that Great Brook Valley has "two kinds of workers – service workers who work at the Marriott and Sheraton [downtown] and semi-skilled manufacturing workers [who work in the adjacent industrial park]." Employers are more than passive beneficiaries of this spatial fragmentation, and one employer in Main South noted how the concentration of similar employers in the area "holds the [Latino] population together" and attracts still more Latino men and women to the area (see also Scott 1988). One way in which employers actively contain local labor markets is through their recruitment strategies and the standards by which they judge potential employees.

EMPLOYERS' RECRUITMENT STRATEGIES

In their recruitment of employees, most employers ensure that the bulk of their work force will not be drawn from across the entire metropolitan region, even one as relatively small as Worcester, only 20 miles in radius. When asked if they had any preferences about where their workers came from, most employers said they did not, but many then hastened to add that they did not want workers who lived too far away from the workplaces. Typical of such comments are the following: "The closer the better; they're happier; there's less stress and strain. They can be relied upon more to stay late and come in early. I think they're more productive if they live closer." From another employer we heard: "I want my workers to live within a certain radius. If I hire a guy from Boston, every snowstorm, he'll be gone."[2]

In addition to their own preference for worker proximity, employers have expectations about the commute times and distances that are acceptable to their workers, and they have incorporated these times/distances into their hiring standards. We were told by one employer, for example, that: "The people we employ won't go more than five miles. There are only two males [out of a labor force of 18] on the premises. Most are married women working for second incomes; a lot work during school hours. Fifteen miles is like global exploration to these people." Employers, especially those in Main South and the Blackstone Valley, were aware that, on average, women's journeys to work are shorter than men's and that production workers are likely to travel shorter distances than are professionals and managers.

Employers' perceptions about workers' travel behavior are quite accurate, in part because employers' preferences and expectations about "appropriate" home work distances have been built into their hiring practices (Figure 4.2 p. 103). Average travel times and distances to work vary significantly by both gender and area, with women in Main South, Upper Burncoat, and the Blackstone Valley spending little time and traveling relatively short distances to employment. Because almost

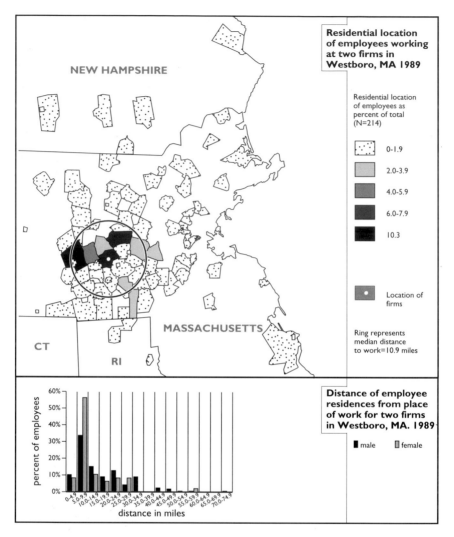

Figure 6.1 (top) Residential location by zip code of employees working in two firms in Westborough, 1989; *n* = 214. (bottom) Distance of employee residences from place of work, Westborough firms

Source: Hanson and Pratt 1992. Record of employees' addresses provided by employer, 1989

all women (94 percent of the employees interviewed in 1989) in Westborough have access to a car[3] and lived in a community that is the site of two major highway interchanges, Westborough women, many of whom are middle class, are able to take jobs farther afield than those taken by women in other areas. In the other areas, women work close to home, much closer than do the men.

The segmentation and localization of labor markets can also be seen by looking at the residential locations of employees of individual firms. Figures 6.1 to 6.4 show the residential locations by zip code of employees in several of the firms we interviewed. The median home-to-work distance rings inscribed on these maps depict compact and in some

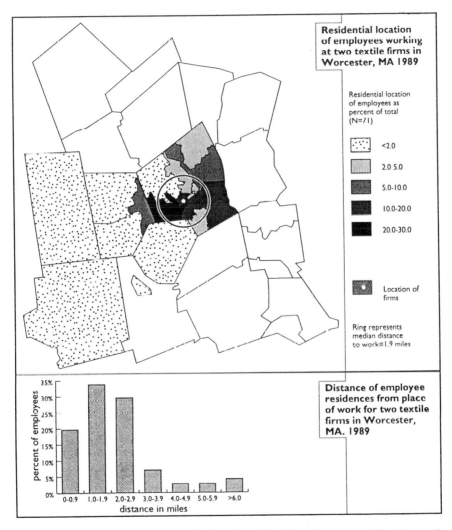

Figure 6.2 (top) Residential location by zip code of employees working in textile firms in Main South neighborhood, 1989; *n* = 71. (bottom) Distance of employee residences from place of work, Main South firms

Source: Hanson and Pratt 1992. Record of employees' addresses provided by employer, 1989

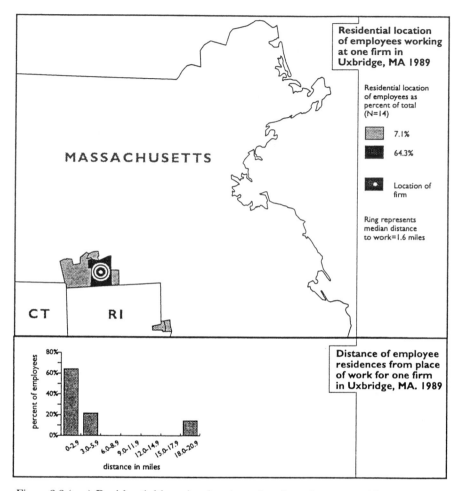

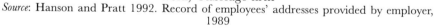

Figure 6.3 (top) Residential location by zip code of employees working at firm in Uxbridge, 1989; *n* = 14. (bottom) Distance of employee residences from place of work, Uxbridge firm
Source: Hanson and Pratt 1992. Record of employees' addresses provided by employer, 1989

cases diminutive labor market areas. Clearly the reach from which firms draw workers varies across the urban landscape, reflecting in large part differences in the type of employment located in different parts of the metropolis. Westborough firms (like those in Figure 6.1, a software developer and a manufacturer of sophisticated heavy machinery with labor forces that are 70 and 80 percent male, respectively) employ large numbers of professional and technical workers, whose residential locations (particularly those of the men) are more widely scattered than are those of people working in the Main South or Blackstone Valley firms. The median home-to-work distance rings for the two textile firms in Main South (Figure 6.2), the yarn manufacturer in Uxbridge (Figure 6.3) and the wool processing firm in Millbury (Figure 6.4) disclose labor market areas tightly bound to the firm's location and covering small fragments of the SMA. While the Main South and Uxbridge firms employ mainly women (with labor forces that are 60, 74, and 73 percent female), the Millbury company, whose workers are

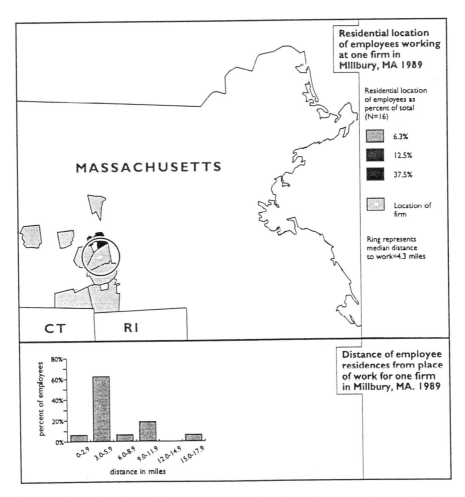

Figure 6.4 (top) Residential location by zip code of employees working at firm in Millbury, 1989; *n* = 16. (bottom) Distance of employee residences from place of work, Millbury firm

Source: Hanson and Pratt 1992. Record of employees' addresses provided by employer, 1989

also snugly clustered around the firm, has a distinctively (94 percent) male work force.

In a minority of cases, employers ensure a local workforce by providing housing. A couple of Blackstone Valley employers continue to provide company housing, which is highly prized by employees for its low cost and proximity to the workplace. Employer-provided housing is not only a remnant from the past; in the tight housing market of the mid- to late-1980s, other companies in the Boston–Worcester area began to extend housing benefits to salaried employees in the form of a shared equity mortgage, offered on the condition that the purchased house be located near the firm (Sit 1991).

Much more pervasive and undoubtedly more influential to the localization of labor markets are the various strategies that employers use to recruit workers. A number of strategies, particularly the use of newspapers and word-of-mouth communication, have the effect (whether intended or not) of localizing labor markets, especially for women (for reasons we explore in the next chapter) and for routine production workers. We asked employers about their use of a variety

Table 6.3 Percentage of employers in each study area that used each method of recruiting labor

	All areas	Main South	Upper Burncoat	Black-stone Valley	Westborough	p-value
Newspaper ads	92.3	88.9	87.5	97.1	93.0	n.s.
Private employment agencies	44.6	26.8	46.7	50.0	56.8	.04
State employment agencies	43.1	51.3	62.5	25.9	39.0	.07
Word of mouth	93.9	95.1	87.5	96.8	93.0	n.s.
Recruitment incentives offered to current workers	34.9	20.0	31.3	36.7	48.8	.05
Churches	25.8	22.5	40.0	35.5	16.7	.16
Posting help wanted signs	13.1	17.5	12.5	9.7	11.6	n.s.
In-house advertising	55.0	50.0	56.3	53.3	60.5	n.s.

Source: Personal interviews with Worcester area employers, 1989
Note: The *p*-value is from a χ^2 test comparing the areas; n.s. = not significant

of possible methods of finding workers (Table 6.3). The most common, used by almost all employers in all four areas, were newspaper advertisements and word-of-mouth recruiting.

Despite the fact that almost all employers use newspaper advertisements as a means of recruiting workers, the actual newspapers used and their geographic coverage varies by area and by occupation so as to reinforce existing spatial segregation. Of the firms placing newspaper advertisements for unskilled or clerical workers, almost all (99 percent) used the *Worcester Telegram and Gazette*, but three-quarters also placed ads in other papers, all of which have extremely local and nonoverlapping market areas. When asked in an open-ended way why they used these newspapers to recruit unskilled and clerical workers, 79 percent of employers said that it was to tap a local labor force.

Many Blackstone Valley employers, for instance, place ads for production jobs in the local (Blackstone Valley) paper only; "We use only the local newspapers because people don't want to travel more than ten miles from where they live." A Westborough employer sums up a strategy that many firms follow in recruiting their nonprofessional employees: "I don't use the Boston papers; you get people who live all over the place like the North Shore, so we advertise only in the *Middlesex News* [a local paper whose market area is the suburbs along Rte. 495] and the *Worcester Telegram and Gazette*." The Boston papers are reserved for recruiting skilled production or sales workers, engineers, and professionals. Employers recognize, therefore, a hierarchy of spatial scales in relation to different occupations and target newspapers that they consider appropriate to the labor market reach of particular types of workers. Their own preferences and their assumptions about workers' preferences regarding journey-to-work distances lead employers to place ads for less-skilled jobs and clerical work in very local papers, thereby reinforcing nonoverlapping and local labor market areas among these occupational groupings.

Many employers used newspaper advertising to fill only professional jobs and relied on word of mouth for filling other, nonprofessional ones. Word of mouth was an extremely common recruiting strategy, used by nearly all of the employers (94.3 percent) who were interviewed. Seen as an inexpensive and efficient recruiting method, word of mouth is embraced enthusiastically by firms of all sizes and scopes, not just those with a local orientation. One-quarter of the firms in our sample said they filled at least 80 percent of their jobs through word of mouth; only 10 percent of the firms said they filled less than 5 percent of their jobs this way.

Employers were drawn to this strategy for a number of reasons. A minority of employers used it as a way of gaining access to a particular ethnic enclave, recognizing, for example, that workers who knew each

169

other could share the expenses of a car and commute together to work. One Upper Burncoat manufacturer hired exclusively Puerto Rican male production workers, all living in Great Brook Valley. He relied entirely on a rather passive, but nonetheless successful, word-of-mouth recruitment strategy:

> They never give notice; they just quit. When they leave, they never come back. But they always fill the job with someone else, always another Puerto Rican. It's just their culture. They have a different protocol. They know if they bring in people who don't work, it is a real black mark for them. They have been dishonored.

Several employers in Main South mentioned the role of a labor broker as a gatekeeper to Asian workers. One temporary labor office in Main South reported dealing with four brokers, who provided 70 percent of their Asian workers: "Once you've got a broker (actually, it's more likely that they find you), the broker will find you the ethnic workers." This temporary labor office paid the brokers bi-weekly on the understanding that the broker would then pay some percentage to the workers: "Brokers give us a family package, mostly Asian. If the son can't make it, the mom fills in." The brokers maintained considerable control over this labor supply, in part because of the inability of many workers to speak English. One employer in Main South tried to hire on a full-time, regular basis the Vietnamese workers whom he had been employing through a temporary agency:

> The temporary agency deal is that after 12 weeks you can hire them permanently. We wanted to hire them all permanently because they are such good workers . . . but those leaders control them. We had a meeting to tell them about the benefits [of permanent work] and so on. The leader of the Vietnamese temporary agency acted as the translator, and we think he probably discouraged them from being hired permanently because so few (only 6 out of 18) opted to join us full time. We pay them all in one check. I think he pays them in cash from that. [The employer was certain that each worker had a valid green card.]

Other employers mentioned the difficulty of assessing prospective employees in what they felt to be an increasingly uncertain and dishonest world. One employer noted: "Their track record, their job history . . . it's increasingly hard to check on that sort of thing. People lie on their employment forms." An owner of a bonded warehouse company also indicated his inability to judge the honesty of applicants recruited through newspapers: "There's a lot of theft when you go through the newspaper."

The majority of employers who used word-of-mouth recruiting saw it

170

as a way of creating a stable, disciplined, productive, congenial, and homogeneous workforce: "If an employee recommends a person for a job, that person generally has the same characteristics as the employee." A large Blackstone Valley abrasives manufacturer explains how this works:

> We've got five sets of brothers; we've got cousins and in-laws. We get about 75 percent of our workers through word of mouth. I prefer it this way because people work better together who know each other, and you've got production incentive. They don't want someone in here who's going to drag his feet. Most people only want good workers in here; it's in their interest to recruit other good workers.

A predilection for recruiting through word of mouth is not limited to manufacturers seeking production workers; a software developer and marketer in Westborough expressed a similar philosophy about hiring through personal contacts: "It's the best recruiting strategy. There are two considerations when we hire someone – first is competence, and second and more elusive is work ethic and how well someone fits into the culture of the company."

Several employers also mentioned that they had a policy of hiring employees' children during the summer months, as replacement workers for those on vacation. While this policy extends the likelihood of family networks among the employees, one employer noted that the policy had more to do with promoting good will among existing employees than screening for potential new employees:

> The kids don't usually come back to work permanently, but it gives the parents a chance to know where their kids are . . . mother works here and daughter works over there. It gives the kids the chance to make money and work in a nice plant. Parents feel good about it. It creates good feeling.

Yet another reason employers favor recruiting through their workers' personal contacts is to promote the smooth transmission of skills on the job (Manwaring 1984); workers are more likely to provide efficient, informal, on-the-job training to people they have brought on board (people who are "like themselves") than they are to strangers.[4] Several scholars have drawn links between employers' recruitment methods on the one hand, and employees' skills acquisition and promotability, on the other (e.g., Corcoran et al. 1980; Granovetter and Tilly 1988). Barron et al. (1985) note that employers are likely to put more effort into actively recruiting employees for jobs where formal (educational) qualifications are important and where formal on-the-job training is extensive. In a survey of Worcester-area employers, Budz (1978) detected a casual attitude among employers toward the on-the-job training and advancement of women; she saw this attitude as an important reason behind employers' tendency to hire women "walk-ins" rather than pursuing more formal recruitment strategies for hiring women.

The more general point is that employers' recruitment methods can intersect with on-the-job training possibilities in ways that lock some workers out of job advancement. As an example of this, one employer with whom we spoke bemoaned the inadequate training that young people currently receive in trade school and the resultant need for his workers in one job in particular (press operators) to learn the job entirely on site: "The local trade schools don't offer the kind of training we need anymore. We've been caught flat-footed without a trained replacement for press operators a few times. No one comes in here knowing how to operate these presses." The fact that all the press operators in the firm were men (although half of all the workers in the establishment were women, including a number of production workers) suggests that women are excluded from this in-house training, which would give them access to the $11–$12-an-hour press operator job instead of the $6.50-per-hour production jobs women held. This employer also reported using word of mouth to recruit production workers, although we have no way of knowing if the male press operators specifically were recruited this way.

Whatever the reasons for using word of mouth, the strategy tends to reproduce existing gender- and race-based occupational segregation. Particularly in enterprises where employers rely upon employees' personal contacts to recruit new workers, where on-the-job learning is important to upward mobility, and where much of this learning takes place informally through personal contacts on the job, segmentation of the labor market is likely to remain entrenched. As Stevens (1978) notes, even when there is no overt discrimination on the part of employers, the use of word of mouth tends to perpetuate racially segregated labor markets, as individuals in particular groups circulate and contain information about job opportunities within their group. Given the striking gender segregation of job information networks, discussed more fully in Chapter 7, word-of-mouth recruitment also has the effect, however unintended, of perpetuating sex-based occupational segregation.

This manner of recruiting labor also has implications for segmenting the labor market spatially, as well as socially. Using the personal contacts – friends, neighbors, kin, previous coworkers – of present employees to recruit new workers promotes the spatial clustering of employees' residences and strengthens workplace–neighborhood linkages. These neighborhoods need not necessarily be located in the immediate vicinity of the workplace. Mier and Giloth (1985), for example, find that word-of-mouth recruiting explains the seeming paradox of high unemployment and long commutes to low-wage jobs among Mexican-Americans living in a neighborhood in Chicago that had good local employment opportunities. They found that the neighborhood industries used primarily word of mouth among existing employees to fill job openings. Because the existing employees did not live locally and had no local networks, residents living close to these industries were effectively excluded from employment in their neighborhood firms.

We, too, have found examples of established linkages between residential communities and workplaces that are spatially discontinuous. An employer in the Blackstone Valley, for example, had developed relations with a Polish parish in the city of Worcester so as to obtain access to a Polish immigrant labor force. Living within the same neighborhood, the recruited Polish workers could travel together in one car. In general, however, we have found that employees tend to be drawn from local neighborhoods (Figures 6.1–6.4) and, thus, the use of word of mouth to fill job openings is likely to yield a highly localized workforce.

EMPLOYERS' COMMITMENT TO PLACE

Once employers are situated in a particular place, in part because of the local labor pool, in many cases they become tied, not just to Worcester, but to a specific part of the urban area defined by their current labor shed. Contemporary academic and popular accounts tend to emphasize the international mobility of capitalists, spurred in part by a search for cheaper forms of labor. This no doubt is happening, and we find ample evidence of it within our case studies: for example, one large electronics firm in Upper Burncoat had moved their assembly operation in the late 1960s to Mexico and then, as labor costs and union activity increased in Mexico, to the Philippines. Further, the residents of the Blackstone Valley have witnessed a fitful but steady exodus of textile manufacturers since the 1920s, many leaving the Valley in search of cheaper labor. Plate 6.1 captures both ends of this spatial continuum. In this 1977 advertisement, the Norton Company mentions its multiple production sites while presenting Norton and Worcester, company and place, as synonymous. Underlying the importance of local recruitment strategies, this advertisement was placed in the yearbook of a nearby high school (Burncoat High).[5]

Many employers expressed extreme geographic rootedness. In a straightforward (and perhaps naive) question we asked employers point blank if they felt rooted to the Worcester area; 90 percent said yes. Moreover, large establishments or those that were parts of multinational firms were no less likely to say they felt rooted than were smaller or purely local firms. Half of those who felt rooted to Worcester attributed

Plate 6.1 The Norton Company's advertisement in a local high school yearbook
was one means of recruiting a localized labor force
Source: Burncoat Senior High Yearbook, Worcester, Mass. 1977, p. 158

this to the skills, dependability, experience, loyalty, or productivity of
their current labor force. Many employers were conscious of the invest-
ments they had made in training their current labor force and noted an

174

unwillingness to risk losing their present employees by moving, even to another area within Worcester. Two Main South employers, for example, expressed concern that they would lose their Vietnamese employees, who lived in Main South and relied on public transportation, if they were to move to a preferred suburban location. In one case, the owner had responded to his workers' complaints about the projected transportation problems associated with the proposed move by deciding against the move. In the other, the owner had decided to move his envelope manufacturing facility to a suburban location, hoping that the Vietnamese workers, unhappy with the move, would nonetheless purchase cars and continue to work for his firm at the new location.[6]

About two-fifths (44 percent) of the firms we interviewed had, in fact, changed location within the past five years, but practically all had intentionally moved only a short distance[7] so that the bulk of their employees from the previous location would not be lost to the firm. Several employers, forced to relocate to larger facilities, actually mapped the residences of their current employees, in an attempt to find the least disruptive new location. A worker at a wire production facility that moved from one location in Main South to another in 1982 explained that:

> One of the reasons the owners decided on this new location is because they drew a map of where everybody lived geographically. They saw where they could move and still have everybody come. All of the people here are at least semi-skilled and it would be bad to lose them.

An owner of a factory in the Blackstone Valley that employed 55 people – 25 men and 30 women – to produce high-speed ribbon cable for the computer, banking, and defense industries tells of a similar mapping exercise:

> I'd rather train people myself. I want smart people, not trained people. . . . Maintaining our current employees would be the primary factor in choosing another location. Because of them, we'd consider places within a 12-mile radius of Millbury because this would be a central location to our employees.

The knowledge of the 12-mile radius had been obtained by mapping current employees' residences; the employer had literally put a map on the wall and placed a pin on it for each employee's residential location.

The strategy of short-distance relocation produces the desired effect: of establishments that had moved, 92 percent reported that most of their workers stayed with the firm after the move, and 86 percent said that *all* of their employees moved with them. These findings are obviously biased by the fact that firms that had moved outside the Worcester region (or even outside our case study areas) were excluded from our study. Nevertheless, three-quarters of all interviewed firms (not just those that had moved recently) said that if they were to move, access to labor would be important to the choice of new location, and two-thirds said that if they were to move, access to their current labor or to workers with the skills of their current labor force would be important.[8] Perhaps one reason for employers' sensitivity to keeping their current labor is, as we noted in Chapter 2, that Worcester has historically been and remains a nonunion, and even anti-union, town; only 10 percent of the firms in our sample had any unionized workers.

Employers are also tied to place through their own personal connections (see Cox and Mair 1988); we encountered an extremely localized group of managers. In half of the firms we interviewed our contact person was the owner, and 87 percent of this group said they were from the Worcester area.[9] Even more striking, of the persons we interviewed in all firms – all of whom were either owners or senior management in influential decision-making positions – 67 percent were from the Worcester area. Personal ties intersect with the recruitment strategies to yield a deep concern for the local area. The manager of a Blackstone Valley printing supplies manufacturer, when asked if the firm had any preference about where their workers lived, said: "I don't think there is any real preference, but we tend to skew toward the locals [in hiring]. The owner was brought up here and wants to support the local area. He lives in [a nearby town] and knows a lot of people." These findings about the importance of employers' family ties are interesting in light of Howland's (1988: 154) finding that single-location firms had a lower rate of closure than did branch plants; she speculates that because of their ties to the local area, owners of single-location firms might be more willing to accept lower rates of profit.

Another factor contributing to some firms' commitment to a particular location is linkage to other firms, establishments that are located in the same area perhaps to tap the same labor pool. Scott (1988, 1992a) has called attention to the increasing importance of interfirm linkages in creating industrial agglomeration effects, in drawing out distinctive pools of labor, and in making firms less footloose. Ties to subcontractors and service suppliers tended to be the most localized kind of linkages (Table 6.4), and proximity to subcontractors was especially prized: about half (51 percent) of the firms we interviewed reported using subcontractors, and 58 percent of these deemed proximity to subcontractors important or very important. Indeed, the bulk of the firms' subcontractors and service suppliers were located within the Worcester region, with a high proportion lying within close proximity of the firm (Table 6.4).

The majority of employers recognized the benefits accruing to them from being part of a local agglomeration of similar employers – an

Table 6.4 Average percentage of firms' material suppliers, subcontractors, service suppliers, and sales that come fron each geographic region

Geographic region	Material suppliers	Subcontractors	Service suppliers	Sales
Local	16 (28)	39 (40)	43 (43)	13 (25)
Worcester region	29 (31)	59 (41)	66 (37)	29 (35)
Massachusetts	44 (35)	73 (40)	78 (36)	48 (39)
New England	57 (36)	69 (45)	77 (38)	60 (40)
Elsewhere	41 (36)	9 (24)	7 (21)	43 (42)

Source: Hanson and Pratt, 1992; personal interviews with Worcester-area employers, 1989
Note: Local area refers to the study area (Main South, Upper Burncoat, the Blackstone Valley, or Westborough). Numbers in parentheses are standard deviations

agglomeration that had a symbiotic relationship with a particular labor force, be it engineers, skilled machinists, Latino workers, or low-waged married women. They saw that their location provided the benefits of a shared pool of skilled (or cheap) labor. Almost all of the employers (90 percent) indicated that they shared a labor pool with other local employers and a minority of this group (34 percent) saw this as a disadvantage. Given the existence of specialized employment districts,

Table 6.5 Percentage of employers sharing each type of labor with other employers in the study area

	Main South	Upper Burncoat	Blackstone Valley	Westborough
General/unskilled (e.g. janitors)	10.8	0	18.5	15.0
Routine production	67.7	50.0	70.4	27.5
Skilled production	54.1	25.0	51.9	25.0
Engineers	0	25.0	0	27.5
Clerical/office	5.4	16.7	11.1	32.5
Sales/marketing	5.4	16.7	7.4	12.5
Professonal/managerial	2.7	8.3	0	12.5

Source: Personal interviews with Worcester area employers, 1989

it is not surprising that employers in different areas share different types of workers (Table 6.5). Those who found it disadvantageous to share labor with other local employers tended to be sharing unskilled rather than skilled workers; the labor market was so tight for unskilled workers at the time of the survey that one employer reported having bailed people out of prison to obtain workers, and managers of fast food chains were discussing career incentives to retain low-level personnel (Herwitz 1987: 54).

Most of the employers interviewed, then, recognized that their current location provided a particular conjunction of linkages to other firms, to a labor force with distinctive and desirable characteristics, or to their own personal networks of family and friends. Because of their own long-standing ties to the Worcester area, employers' mental maps of the labor landscape within the metropolitan area are sharply developed. They drew upon this local knowledge about spatial variations in labor characteristics when selecting locations for their establishments. Employers use recruiting strategies that target local labor, and they evaluate potential workers with respect to the characteristics of this local labor force that they have chosen by dint of their location. They go on to reproduce some of these features, for example, through in-house training. The availability of a trained and known labor force proceeds to bind many employers to very circumscribed places, for example, to those within a certain radius of their current location or to a particular public transportation route.

These relations between employer strategies and local labor supply build toward a geography of labor market segmentation, and this geography contributes to class-, race- and gender-based occupational segregation. The existence of extremely localized labor markets, each associated with a distinctive employment district and offering different mixes of occupational opportunities, suggests that access to particular types of jobs is limited to those who live in particular parts of the metropolis. A woman living in the Blackstone Valley, for instance, is unlikely to know about a clerical job opening in Westborough because, even if advertised, the opening will most likely be advertised only in the local Westborough newspaper. More likely, the job will be filled through word of mouth, the information about it filtered through social networks of existing employees, networks that are often local and gender-, race-, and class-specific.

Thus far we have focused on the containment and separation of different intraurban labor markets, such that those living in one part of the city develop a special relationship with the local employers. This special relationship both fosters the development of particular arrange-ments and skills, and makes employees less likely to gain access to firms outside of their local community. Shifting our perspective to the

178

individual firm, we can see how geography further structures occupational segregation within firms.

GEOGRAPHY AND OCCUPATIONAL SEGREGATION WITIIIN FIRMS

There are two ways that spatial relations sharpen and perpetuate differences among those who work within the same firm. First, and at least partially following from the employers' recruitment strategies already discussed but also owing to class and race-based residential segregation, those who work in different occupations within the same firm tend to come from different parts of the city. In Figure 6.5 we have plotted the residential locations of employees in different positions in a large insurance company. Not only are the residential locations of managers (84 percent of whom are male) more widely dispersed than those of actuaries (78 percent of whom are men) and clerical workers (84 percent women), the residential distributions of the different occupational groups vary somewhat; the locations of clerical workers' residences, for example, differ from those of actuaries, with more clerical workers living in south Worcester. These patterns develop, at least in part, because employers use different channels to advertise for varying types of workers, being more likely to use local newspapers to fill unskilled and semi-skilled production and clerical jobs and using metropolitan and national papers for some skilled production and managerial and professional positions. Employers' propensity to recruit through employees' social networks also reinforces the tendency to attract a particular category of worker from the same residential area.

The outcome is that differences between groups of workers and the status hierarchy within the firm are reinforced by different residential locations. As one example, a male Latino production worker at an industrial plating firm in Upper Burncoat outlined a detailed social geography of workers' residences, broken down by occupational type and gender. All of the male production workers come from the adjacent public housing project. He mentioned, however, that the female secretary comes from Shrewsbury, a middle-class suburb that lies directly on the other side of the industrial park in which the firm is located. When mentioning Shrewsbury he added, "That area, you know," simultaneously making a face to signify what he perceived as the snobbishness of those who live there and the extreme social distance between himself and Shrewsbury residents despite the close geographic proximity.

The spatial organization of workers within firms tends to segregate different groups of workers so that, in many instances, they rarely see, communicate, or socialize with one another across occupational (or

179

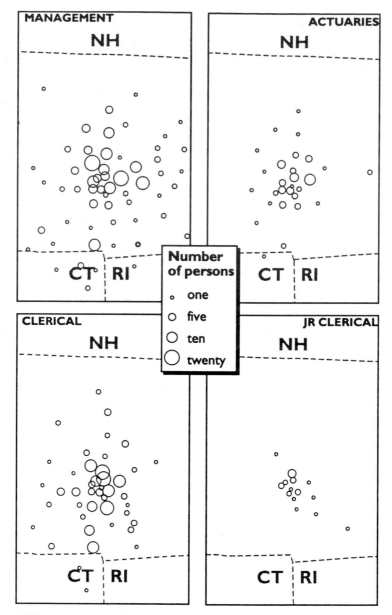

Figure 6.5 Location of employees' residences for an Upper Burncoat insurance
firm, broken down by occupation type
Source: Record of employees' addresses provided by employer, 1989

gender) boundaries. The 1910 photograph of stitchers at the Worcester
Corset Factory illustrates the historical depth of this point (Plate 6.2). In
many workplaces, workers still have very little knowledge of what goes on

Plate 6.2 This 1910 photograph of corset stitchers at the Royal Worcester Corset Company illustrates the spatial segregation of workers by gender and occupation within the firm. Photograph from the Worcester Historical Museum, reprinted with permission

beyond the confines of their own work area, little knowledge of the type of work done by others in the firm. Office work is usually spatially segregated from production work. Breaks are taken at different times and in different places.

In the 1989 interviews with employees in manufacturing and producer services firms we asked what proportion of the person's immediate co-workers (in the same work area, department, or team) were women (men). For women, the most common answer (accounting for 28 percent of all responses) was that *all* immediate co-workers were women. This finding is very much in line with the studies Spain reviews of workers' contacts with co-workers of the same (or different) sex on the job (Spain 1992: 209–11); she reports that between 20 and 25 percent of workers "have no contact of any kind with the opposite sex in the workplace" (Spain 1992: 210). Divisions by gender are often welded to divisions by occupation that are literally mapped into the spatial divisions of the work process within firms.[10]

An employee in an industrial laundry studied in Main South, for example, detailed the work process there. The white male population tended to be spatially segregated from women and immigrant men. The laundry was brought in from hospitals by white Anglo male truck drivers, and then hand-sorted by Jewish and Polish men. It was then moved through the washer and dryer by a conveyor-belt and robots, monitored by white male workers. The male general laborers transported the cleaned laundry to an ironing room where mostly women ironed it. It was then moved to a folding room, where it was folded by hand by women, and taken to a shipping area, where male and female packers boxed it. A woman at a footwear factory in Main South told us: "I've worked here for ten years but I've never been upstairs. That's where the men are."

These spatial divisions within the workplace can make it difficult for workers to learn new skills on the job or to see a path of upward job mobility within the firm. A woman who worked in barring (a "woman's" production job) at a firm in the Blackstone Valley had a sense of how the spatial segregation of work tasks limited both her accessibility to other occupations and her understanding of the productions process: "We try and get them to take us on a tour so that we know what we're doing, what we're making."

The spatial segmentation of the work process within workplaces no doubt both reflects and exacerbates the gendered nature of social networks and the way these networks are inserted into the recruitment process (discussed in more detail in Chapter 7). Even when work tasks are not spatially segregated, gender and spatial segregation of socializing may perpetuate separate networks of friendship and information. At a mailing service in Main South strict gender segregation of occupations was reinforced by patterns of socialization: "Everyone comes into the break room all the time, except for the guys. They eat at their machines or outside."

Spatial and occupational differences also underline differences between various categories of women workers. A woman who works in customer services at a newspaper supply company in the Blackstone Valley claimed: "I don't know anything about production; I don't even know what it looks like out there." She eats her lunch with others in the clerical and customer service departments but "never sees" workers in production and sales (among whom there are a good number of women). As for management, "Well management, you see them around all the time, but you don't have a conversation with them." Her patterns of communication almost exactly replicate the spatial organization of the firm: the customer services and office staff are located downstairs in a room of small cubicles created by room dividers. Sales staff and management are upstairs, in shared or private offices. Production occurs in an entirely separate part of the building, with individual production departments separated on the basis of machinery and work task. Among the production workers, there is very little communication across departments, members of each department meet daily for coffee breaks and lunch, and workers in each department have their own separate annual parties. An employer at a textile mill in the Blackstone Valley captured the difference among women who worked in production and office jobs in the most appalling terms. "It's strange country up here," he told us. "All the women are the same girls in high school. Five years down, they are in the shop and they are changed. They look haggard, old, they don't want to come up front [into the office area]."

It is worth considering the gulf between the women who work in clerical and production work. One woman who worked at the newspaper supply company described above told us how nervous she was to apply for an office job. With a job history that included working in pizza places and in production jobs at several mills, she decided to apply for an office job:

> I always wanted to do office work, but was a little scared that I couldn't do it. But I knew my cousin did it and she worked here and I could always ask her if I needed to . . . I was intimidated by the well-dressed women in the office and the big telephone. I didn't know if I could do it and I thought they [the office workers] made more money than anybody working in the shop.

She got the office job, and partially through socializing with the women in the office ("We took lunch together. Everybody was real friendly.") developed the confidence that she could do the job. At the time of being interviewed, she had been moved out of the front office and was working in shipping. She much preferred this because, contrary to her initial assumptions about office work, she made more money as a shipping clerk.

Certainly spatial segregation is not the only factor conditioning occupational segregation and perceived differences among workers. It is, however, one element that breeds strangeness across groups of workers and boundaries of social differences within firms, while simultaneously building cohesion within localized social groups.

CONCLUSION

We have sketched how employers' spatial strategies, in relationship with existing spatial differentiation of social groups and industrial districts, perpetuate place-based social differences and a geography of labor market segmentation, at two spatial scales – inside the firm and among different areas of the city. Inside the firm, occupational segregation and labor segmentation has a spatial expression such that workers in different occupations and segments are unlikely to exchange information or even see each other. At the neighborhood scale, different occupational and industrial districts develop in part because employers (with preconceptions about labor market skills, wage demands, and work attitudes of different social groups) locate their establishments so as to gain access to specific types of employees. Localized recruitment strategies, through newspaper and word of mouth, effectively build semi-permeable walls around these districts so that local residents typically have the best access to these jobs. Employers' reliance on their employees' social networks for filling job vacancies can lead not only to geographic segmentation of the labor market at the neighborhood scale but also to spatial segregation of workers within the firm, with important implications for a worker's ability to learn new skills on the job.

7

COMMUNITIES, WORK, AND GENDER RELATIONS

"I woke up one day and said, 'I'm going to work'. I just went up to the mill and asked, and they put me on that day." This is how one woman we interviewed in the Blackstone Valley described how she found her job – a job as a comb winder that she worked at for the next 21 years. The mill was located next to her house, a house that she had inherited from her parents. She chose to work in the mill because the only other options that she saw as being available locally were working as a cashier in a retail store or as a waitress, and she had no desire to work "with the public." She knew about the likelihood of a job at the mill because two of her friends worked there. When she started, her hours were from 7:00 a.m. to 3:00 p.m. in the afternoon, times when her husband, who worked the 5:00 p.m. to 1:00 a.m. shift, could look after their eight children. All of her children later worked at the mill.

This brief account captures much of what we have learned about the importance of local places in Worcester for structuring opportunities and decisions that create distinctive patterns of work and ways of life. What is judged to be possible and what is actually available as employment depends on the local place. Our knowledge of places and opportunities is learned and understood through social relationships, which for many are also rooted in place. In his open-ended interviews with 30 young men living in three Boston neighborhoods, Wial (1991) learned that what is considered a "good" job and what is considered "appropriate" work for (in the context of his study) adult men are defined locally. The casual acquisition of skills (such as those related to house and car repair), as well as job search methods, personal contacts, job information, and, in fact, the gendering of jobs are all likely to be dependent on the local context. In Fernandez Kelly's words (1994: 99, 101), "Learnings originate in circumscribed spaces"; "This is what I mean by *embodied* knowledge."

When we began to write about women and work in Worcester, we stressed the homogeneity of women's work experiences, and the pervasive tendency for women to work in low-paying, female-dominated occupations. We laid the sameness of women's work experiences against a far

185

richer residential geography, one patterned by differences of family type and occupational standing (Pratt and Hanson 1988). We are now attuned to a more complex layering of geographies and see a multitude of differences across communities located within the same metropolitan area, in terms not only of the occupational standing and family-type characteristics of local residents, but also locally available services and employment resources. These differences create different frameworks within which families construct their lives and experience gender. There are many ways of living gender, and there is considerable variety in the ways that female-dominated jobs are experienced. This variability is structured in and through places. Feminist theory stresses that the acquisition of gender identity is a political process; to this we add that it is a geographical one: it is at least partially structured by local context.

In our assessment of locally acquired gender identities, we are most attentive to occupational and class differences among various neighborhoods. We make no claims to exhaustion; certainly other experiences structure gendered identities along different lines (for example, the experience of being part of a lesbian community). Our theoretical predispositions and the methods that we have chosen have led us to abstract and highlight occupational and class differences across communities. Our argument, which builds from our discussion of employer practices in the previous chapter, is that varying class and gender practices are created and re-created simultaneously in different urban neighborhoods. Residents in different places attract different clusters of employers, partially because of place-based cultures of class and gender. The employment districts that develop then offer variable opportunities and impose different constraints, which influence the ways that class and gender are lived in these communities and are part of the way that some of the differences among women are produced. Residential communities, particularly ones where residents are rooted through networks of friends and kin, also condition residents' expectations of what is appropriate for men and women. They are contexts in which people develop "morally adequate accounts" (Jordan *et al.* 1992) and expectations about gender-appropriate behavior, for example, whether and how women should combine mothering and paid employment.

To argue that gender is conditioned by community-based class experiences is to highlight the importance of both class *and* geography. As other geographers (e.g., Massey 1984; Sheppard and Barnes 1990; Thrift and Williams 1987; Walker 1985) have taken pains to point out, geography is inseparable from class formation; classes are not just distributed in space, they are constructed in and through space. In the words of Thrift and Williams:

> clearly space has a lot to do with how different people are constituted differently. Institutions are not equally distributed in space, and in

particular locations the prevalent *mix* of institutions will be more or less effective in binding particular people's consciousness [and practices] in certain directions rather than others.

(Thrift and Williams 1987: 16)

Class formation and consciousness are sociological categories that must be thought of in relation to particular historical geographies because they are shaped through those contexts. So too, the political process of acquiring a gendered identity is conditioned by, and in turn conditions, local institutions and resources, some of which are tied to localized class practices.

We begin by revisiting the residential rootedness of households in Worcester and by considering the impact of such stability on social relations, particularly the growth of social networks in place. We examine how, for many households, the areas that are considered as possible places to live are defined by the household's community contacts. We also consider how, for most women and in many communities, the employment relation emerges out of these networks of locally based associations. We then explore how employment and services vary from neighborhood to neighborhood in ways that affect women's occupations, the organization of work within households, and ultimately gender relations.

STAYING PUT

As we saw in Chapter 5, Worcester is a city that holds its residents: many are loathe to leave. As Michael Brown, one of our research assistants and himself a long-term resident of Worcester, pointed out to us, this culture of rootedness, of staying put, is celebrated in the obituary column of the *Worcester Telegram and Gazette*. Pursuing his observation proved to be an enlightening excursion into Worcester's local culture and is worth recounting. Scrutiny of 14 days' obituaries in the *Worcester Telegram and Gazette* (for 203 people) revealed that the person's connectedness to the Worcester area – not only how long he or she had lived here but also often details of the person's residential history – was mentioned in two-thirds of the write-ups. For comparison, we examined several weeks' worth of obituaries from two other small New England cities, Rutland, Vermont and Burlington, Vermont. In the *Rutland Daily Herald* connectedness to place was reported in only one-third of the cases and in the *Burlington Free Press* it was included in only 19 percent. The obituaries in the Worcester paper suggest, moreover, that long-term residence in the Worcester area was highly valued. Considering only those who had been living within the Worcester MSA at the time of their death (the *Worcester Telegram and Gazette*'s market area extends to all of Worcester

Marion Kesseli, 88

WORCESTER — Marion (Turner) Kesseli, 88, of 361 May St. died Thursday at home.

Her husband, Herbert J. Kesseli, died in 1961. She leaves four sons, Robert T. Kesseli of Auburn, Carl H. Kesseli of Worcester, Philip B. Kesseli of Jefferson and John H. Kesseli of Gorham, Maine; 12 grandchildren; and 12 great-grandchildren. She was born in Worcester, daughter of Joseph H. and Beatrice S. (Holden) Turner, and lived here all her life. She graduated from North High School.

Mrs. Kesseli

Mrs. Kesseli was a longtime active member of the Worcester YMCA and its World Service Committee. She was a longtime board director of Hahnemann Hospital, and past president of its Auxiliary. She was a longtime board director of the Worcester Fresh Air Fund at Camp Putnam in New Braintree. She was a member of the Worcester County Music Association and a charter member of Chapter R, PEO Sisterhood.

Funeral services will be held at 11 a.m. Wednesday in First Baptist Church, 111 Park Ave. Burial will be in Hope Cemetery. Calling hours at Caswell-King Funeral Home, 474 Grove St., are 4 to 7 p.m. tomorrow. In lieu of flowers, memorial contributions may be made to the Worcester Fresh Air Fund, Rutherford Road, Camp Putnam, New Braintree 01531.

Plate 7.1 Obituaries in the *Worcester Telegram and Gazette* stress residential rootedness and ties to the Worcester area
Source: *Worcester Telegram and Gazette*, reprinted with permission

County), we were able to discern from the obituary that 70 percent of the people had lived in the Worcester metropolitan area for at least 40 years. Typical examples: "She was born in Bic, Quebec and lived here [Worcester] 84 years"; "She was born in Hubbardston and lived here [Worcester] most of her life"; "He was born in Worcester, . . . and lived here most of his life. He then lived several years in Holyoke and 11 years in Chicopee." Perhaps the most typical is the following. "She was born in Worcester, . . . and lived here all her life" (Plate 7.1).

Curious about the source of this attention to rootedness, we learned that the information included in the *Worcester Telegram and Gazette* obituaries is supplied to the paper by funeral homes. Conversations with 25 funeral home directors revealed that they included information about connectedness to place because relatives of the deceased thought such information was important and considered it essential to a person's identity: "It identifies people. It helps those reading the obituary to know exactly who has died." The funeral home directors also believed that people in Worcester wanted to know about a person's connectedness to this place: "It is valued here."

As we have seen, the transfer of residential property from older generations to younger ones not only grows out of, but also helps to perpetuate, these enduring ties to place. In this chapter, we expand our discussion of rootedness by narrowing the spatial focus to particular communities. People's bonds to place are not just to the Worcester area in general but to certain small areas within it, and these bonds shape the geographics of employment in significant ways.

One measure of such rootedness to a very specific place is how long people have lived in their present homes. Half of the households in our sample had lived in their current residences for at least five years, and two-thirds had been in the same place for at least ten years. It is worth noting that in this sense the Worcester sample is not exceptional; 53 percent of the U.S. population in 1990 was living in the same house or apartment they were living in five years previously (U.S. Census of Population and Housing 1990a).

Moreover, as the residential mobility literature has long reminded us, when people move house, they are much more likely to move over shorter distances than over longer ones. But exactly how short these distances are and the implications of such short moves for rootedness to an area need to be underlined. We asked the households in our Worcester sample who had lived less than ten years in their present home where they had lived before. For all households, the median distance they had moved from their previous to their present homes was 4.3 miles. Among those who had grown up in the Worcester area and had moved, the median length of such residential moves was only 2.9 miles.[1] Clearly, then, when we speak of residential rootedness, we must recognize that people's roots not only have taken hold tenaciously but somewhat diffusely throughout the Worcester area; they are deeply sunk into particular communities within it.

Living in one place for a long time has important implications for how the housing and job markets function because residential stability nourishes the development of personal networks. Insofar as these networks of friends and acquaintances, relatives, neighbors, and co-workers are also tied to particular locations, they in turn foster rootedness. One way that personal contacts do this is by being lively conduits of information about housing and jobs. If the housing and job possibilities that an individual's personal contacts know about are largely confined to a particular area, then finding jobs and housing informally through personal networks will mean narrowing the geographic scope of places where they are likely to be found.

With residential longevity also comes a certain stability in one's daily travel-activity patterns as some places are visited, and routes traversed, repetitively. These places that are encountered in the course of everyday life – stores, restaurants, banks, laundromats – can also be an important source of information about housing and employment opportunities, through, for example, "for sale," "for rent," and "help wanted" signs. Like the knowledge and tips that flow through personal networks, such information is also spatially biased toward the local, toward the areas in which one lives and works.

In fact, finding housing and employment through informal means (either through personal contacts or through one's daily travel-activity pattern) was extremely common in our Worcester sample. Almost half (49 percent) of the households we interviewed had found their current residence this way, and more than three-quarters (77 percent) had found their current jobs this way. Moreover, having grown up in the Worcester area significantly increased one's chances of having found housing or employment informally as opposed to through more formal channels such as newspapers or agencies.[2] That longevity in place yields personal contacts that connect people to housing and jobs is perhaps obvious. That such personal ties tend to have a significant geographic impact by circumscribing one's life space is perhaps less so.

FINDING HOUSING

As we have seen, residential location – where one lives within the Worcester area – is extremely important, especially for women, in defining the employment options that are available. How people find housing becomes, therefore, salient to understanding the geographies of gendered employment.

We asked people a series of questions about how they had gone about locating their housing. In one, we asked whether or not they had looked for housing in any other parts of the Worcester area when they were looking for their current house or apartment. About half (47 percent) of the households said they had not; they had, in fact, looked only within the area of Worcester within which they were living when we interviewed them. But the proportion was far higher among those who had found their housing through informal means. Fully two-thirds of this group, compared to 29 percent of those who had used a realtor, newspaper, or other formal channels, said they had confined their search for housing to the particular part of Worcester in which they were living at the time of the interview ($p = .000$). A clear and strong relationship exists, then, not only between rootedness, on the one hand, and the use of informal channels of information for finding housing on the other, but also between using informal networks and searching for housing in a confined area. This latter relationship no doubt reflects the local intensity of the lived experiences that people communicate through their personal networks. It also helps to explain why, as we have seen, so many households move over such short distances.[3]

People who had grown up in the Worcester area not only relied more heavily than "outsiders" on personal contacts; they also were likely to get information about housing and actually to have found a house or apartment through contacts that were noticeably different in kind from those used by "outsiders" (Table 7.1).[4] Long-term Worcester

Table 7.1 Impact of residential rootedness on types of personal contact used in finding housing

Type of personal contact	Respondents who grew up in Worcester (n = 354)	Respondents who grew up elsewhere (n = 230)	p-value
Family			
Used	57.6	30.4	.00
Found	29.5	15.2	
Neighbor			
Used	16.1	12.6	
Found	3.1	0.9	
Co-worker			
Used	21.3	31.1	.02
Found	3.4	7.0	
Friend			
Used	45.9	45.4	
Found	13.9	17.7	

Source: Personal interviews, Worcester MSA, 1987
Note: "Used" means person obtained housing information from this type of contact (includes cases who found housing through this type of contact). "Found" means person found current house/apartment through this type of contact

residents were far more likely to receive housing information from relatives and far less likely to receive such information from co-workers than were people who had not grown up in Worcester (Table 7.1). Almost one-third (29.5 percent) of those who had grown up in Worcester had actually found their current residence through a family contact, and an additional 28 percent had used information from family members in their housing search.[5]

Not only do relatives play a major role in linking households to housing; they also tend to be implicated in housing searches that are focused on particular locations. Fully half of those who had received housing information or found their current housing through a relative said they had not looked elsewhere within the Worcester area for a place to live. In contrast, relatively fewer of the households who had talked with neighbors (31 percent) or co-workers (26 percent) about housing opportunities had limited their housing search to one particular area. Among those for whom friends were the source of housing information, two-fifths had constrained their searches to one area of Worcester. When we asked people specifically why they had not looked elsewhere but had limited their search to the area they had chosen, 27 percent gave reasons related to either already being rooted to that area themselves or having relatives there. Clearly family members are instrumental in attracting people to particular communities.

One way that family networks work to draw people to certain parts of the urban area is through widespread concern to keep child care within the extended family. One young professional couple had settled in Shrewsbury, an eastern suburb, because, the husband told us, "My wife wants to live near her mother." Not coincidentally, the wife's mother also takes care of their young daughter when needed. The wife, who manages representatives of a computer firm, works at home when she is working locally but travels out of state for a total of about two weeks of every month. This rootedness to Shrewsbury (which for the wife was nonnegotiable) entailed a 35-minute commute for her lawyer husband, a rather long work trip in the Worcester context.[6] A similar example in a working-class suburb in the Blackstone Valley involved a young family moving to be close to the husband's parents, who were to be the family's main source of child care. Because the grandmother did not drive, the family had to find a place within walking distance to her home, severely constraining their search space. The desire to live close to the grandparents also meant extended commutes for the young parents, both of whom worked as technicians at an electronics firm 35 minutes from their new home.

Living in close proximity to relatives is not motivated only by the desire to facilitate family-based child care. It is often the outcome of complex processes that are based quite simply in long-term rootedness to a particular area. The ways in which family and business concerns intermingle to place relatives in the same residential community are illustrated in this example of how a woman, who owns and operates her own dance school, came, with her husband and child, to live near to her parents in a suburb southwest of Worcester. With lifelong roots in that community, the woman had established her business there and then had started looking for a place to live close to her dance studio. Her parents were the ones who found the house she and her husband eventually bought. Although the house had not been for sale at the time, the parents identified it as ideal, and the family then talked the owners into selling it.

As was the case with the intergenerational transfer of residential property, the practice of locating housing through personal contacts does tend to be more widespread among working-class than among middle-class households, and therefore varies geographically. Using the occupation of the male household head as an indicator of household occupational class, only one-third of professional or managerial households, compared to three-fifths of skilled manual and 56 percent of nonskilled manual households had found their housing through informal channels.[7] These class differences translate geographically into distinctions among communities within Worcester. In particular, our four study areas (Main South, Upper Burncoat, Blackstone Valley, and Westborough) were different from each other in the degree to which the people living in each area had used informal means to find housing. Rather high proportions of households in Main South (72 percent) and the Blackstone Valley (61 percent) had found their housing informally compared to Upper Burncoat and Westborough where "only" 41 percent and 24 percent, respectively, had ($p = .000$). These differences reflect different degrees of residential rootedness among these four areas and to some degree, as we outline below, differences in local norms and local cultures.

To sum up, people in Worcester tend to stay put. This long-term residential stability nourishes personal networks of family and friends, neighbors and co-workers and encourages information about things such as jobs and housing to flow through everyday personal interactions. Such information tends, however, to be spatially biased toward the local to the extent that the experiences of one's personal contacts are spatially circumscribed. As a result, information is not only not ubiquitous or homogeneous; it tends to be sticky over space and, in its content, therefore, to be quite place specific. The accumulated impact of residential rootedness is, then, to underscore the importance of the local community in shaping labor market outcomes. We next explore how these place-based information exchanges function to mold the mottled geographies of sex segregation in the labor market.

FINDING JOBS

In contrast to neoclassical economic models of job search, which convey the impression that individuals search for jobs rationally and purposively across a metropolitanwide labor market, we have found that webs of social relations and the accidents of place play key roles in finding jobs. (See Grieco 1987; Jordan *et al.* 1992; Granovetter 1974, 1985, 1986; Morris 1990; Hanson and Pratt 1991; and Wheelock 1990 for extended critiques of attempts by neoclassical economists to link families to labor markets, and in particular of their tendency to ignore the socially embedded nature of job searches.) People in Worcester tend to "fall into" jobs, discovering them through personal networks and chance encounters, often within the local neighborhood (Hanson and Pratt 1991, 1992). "I was walking the baby," we were told by one 39-year-old woman, "It was a real hot day out. I decided to get soda and a juice for the baby and me. [The store where she found her part-time job as a cashier] was right there. I saw a sign that said that they needed part-time help, so I filled out the application and got the job the next day." Over half of women (57 percent) and men (51 percent) in our representative sample of Worcester households fell into their present jobs rather than conducting an active job search, suggesting that chance encounters and informal networks may lead many people to their jobs.[8]

In fact, as we have already noted, more than three-quarters (77 percent) of the people in our representative sample of Worcester households had found their jobs through personal contacts or through their daily activity patterns (e.g., by seeing a help-wanted ad in the window).[9] Moreover, finding jobs informally,[10] rather than through more formal avenues like newspaper ads or employment agencies, was the norm across all occupational groups; it was not confined to those in working-class jobs: 77 percent of men and women in professional/managerial and 70 percent of those in skilled manual work had found their current jobs informally. The 1989 interviews with (a nonrepresentative sample of) employees in producer services and manufacturing firms reinforce the point that social bonds formed outside employment permeate the workplace, linking people to jobs and workplaces to neighborhoods. Four-fifths of the employees we interviewed had at least one friend or relative currently working in the same establishment, and slightly more than half of these workers had known that person before they came to work there.[11] In some cases a friend or relative had worked there earlier but had left during the respondent's job tenure at the firm.[12] Including such people, 35 percent of the employees that we interviewed in 1989 had one or more relative, and 72 percent had at least one friend working at the same firm. These are not social ties that are made and sustained only in the workplace: three-quarters of those with friends or relatives at work now said that they socialized with these people outside work, often in their home communities.

Networks of social associations structure job choice at a number of levels. At the most basic, they define the boundaries of what is imagined to be possible. Whether one sees oneself as an architect or clerk, comb winder or hairdresser, depends on social milieu, which is also a place. Consider the narratives of two women, one 59 years old, the other 28 years old, both working in the customer services department at a yarn mill in the Blackstone Valley.

> I took secretarial courses in high school. I enjoy office work. I did work in a factory at night. It was not my thing. It was assembly work with wires. I knew I wasn't going to get to go to college. My family didn't have that kind of money. The second best option was to take the secretarial route. In 1948 – that was a long time ago – a lot of people couldn't even finish high school. My husband had to go right to work. He was the oldest of nine children. He had to stop school to earn money. You had to get out and work. College was not an option. Everyone went to work when they graduated. It was automatic that everyone went to Whitin Machine Works. You automatically went to them first. A lot of my classmates worked there. I worked on days and nights. . . . When the kids were small I worked nights at Whitin Machine . . . I worked from 5:00–11:00 [at night]. It was good. We were within walking distance, a lot of girls in the neighborhood worked there.

The 28-year-old woman had worked at the mill for 12 years, starting at the age of 16 while she was still at high school.

> My mother works here. She told me there was an opening in the model department. The location was important. I didn't have my license when I started working here. I would walk to school. Then I walked here. Other people from my high school also worked here. My brother worked here. They had more employees then. When I got this job, there were not many places to work in Uxbridge: waitressing, offices, sales, those kinds of jobs did not appeal to me. I was shy. I didn't want to deal with people. Waitressing is a lot of work, and it requires contact with people.

These accounts also demonstrate the concrete ways that social relationships shape job selection: through the sharing of information about specific job opportunities. An important point is that this job information is most often exchanged through everyday encounters with friends, relatives, neighbors, and coworkers rather than through directed, purposeful job search. We have found that different groups of people tend to rely on varying social networks and channels of information, which have distinctive geographies.

The networks are diverse, and the geographies are multiple and complex. We learned, for example, of many networks that drew specific immigrant populations to particular workplaces. A 26-year-old woman from the Dominican Republic, currently living in Main South and working as a folder and ironer in a nearby industrial laundry, gives a sense of the multiple networks and geographies at her workplace alone.[13] She obtained her job through her brother, who is a "work boss" at the laundry and with whom she drives to work. A number of the Latina workers with whom she associates live in the same building, around the corner from her house: "About 12 people who work here live where I live. All my good friends [at the laundry] live near me in the building that I'm talking about. I started working, and then cousins came and friends came, and all started working." She describes another group of Latino workers who travel about 40 minutes to their job: "The Lawrence group are one group. They're all Hispanic. The company used to be located in Lawrence and was moved here, and these people keep coming. [They commute together in a van.]" She identifies another group of workers, Polish men: "The Polish people live farther away [than her friends]. I know they all live together though. They live about 15 or 20 minutes away. All Polish live in a building far away from here."

These Polish workers undoubtedly live in Vernon Hill, a Worcester neighborhood that is home to many Polish immigrants. Networks of association take other Vernon Hill residents in very different directions. An employer interviewed in the Blackstone Valley, who himself was not Polish, had identified Polish immigrants as a desirable labor pool for his wool processing plant:

> We were hiring people who were warm and upright and that's about it. . . . Now we're trying to get a thinking person in our employ and to provide enough incentive to keep the person here. . . . We have explicitly gone after an immigrant labor force. We have an agreement with a Polish parish that we will employ any of their people. We have six Polish people working for us now – they have a good immigrant work ethic. They tend to live in the same area and drive to work together. A person will write home about his work experience. We take pictures of them in their uniforms and they send them home. They're proud; they're doing something that they're respected for. These men leave Poland for two or three years to work, and work they do – seven days, night jobs – and they send money home. When they get back, we generally get a cousin or brother to take his spot.

The employer had identified this labor pool through his sister-in-law, who worked at the Polish church in Vernon Hill. After she drew her brother-in-law's attention to the large number of immigrant men, he went to the church to visit with the priest, at which point, "we made a mutually beneficial agreement. . . . He tells all the Polish immigrants that he can find them jobs, and I put them all to work." This owner has enough commitment to this recruitment strategy that he has learned Polish and runs a small English as a Second Language class at the workplace.

Interviews with workers at this firm confirm the employer's story. A 40-year-old machinist, still on temporary visa, immigrated to Worcester from Poland in 1987 to join his 80-year-old aunt, who has lived in Vernon Hill for 16 years. He spoke no English when he arrived and found his job through someone he met at his church in Vernon Hill, a male friend who also gave him a ride to work for the first two months, until he could afford to buy a car. He has since told four other men from the Polish church about his workplace. They all live close together in Vernon Hill and carpool to work: "I'm the only one with a car."

Despite the real complexities of the geographies of information flows, several generalizations seem possible. First, those who find their jobs through personal contacts or through daily activity patterns tend to find jobs that are located closer to home than are the jobs people find through more formal means, such as newspaper ads, employment agencies, or unions (Table 7.2). This is true for both men and women, but women who find their jobs informally have the shortest commutes of all, significantly shorter than those of men who find their jobs through informal channels. This likely reflects gender differences in personal contacts and the different geographies those gender differences embody.

This observation leads us to a second generalization: women and men obtain their information about jobs from different sources, and the channels of information to which women have access reinforce their

Table 7.2 Travel time to work by method of finding job

Means of finding job	Women			Men			p value
	n	*Mean (in minutes)*	*Standard deviation*	*n*	*Mean (in minutes)*	*Standard deviation*	
Personal/ informal	236	15.3	13.9	134	19.8	14.8	.004
Formal	73	19.4	16.9	36	24.7	18.3	n.s.

Source: Hanson and Pratt 1992; personal interviews, Worcester MSA, 1987
Note: n.s. = not significant. A comparison of the travel times of those using personal/informal means of finding a job vs. those using formal means is significant for women ($p = .04$) and for men ($p = .09$). Personal/informal combines the following responses: personal contact told me about job, recommended me, helped me get a job, hired me; saw ad in window; direct application; I was recruited. Formal combines the following responses: saw ad in newspaper; through union; through employment agency

197

tendency to find paid employment close to home (for a detailed discussion, see Hanson and Pratt 1991). Two characteristics of women's job contacts in particular reinforce an extremely localized job search. First, the personal contacts through which women find paid employment tend more often to be family- and community-related than are those of men (Table 7.3).[14] Second, channels of information are gendered; job information is most likely to come to women from other women and to men from other men (Table 7.3). For women, this is especially true for information obtained from friends and acquaintances, and from work-based and community contacts. For both men and women, the family contacts used in securing a job are more likely to be male.[15] For women, however, the bulk (63 percent) of these male family contacts were spouses and in a literal sense, therefore, they are also community-based. The gender bias in the channels through which job information flows is most striking for men: not a single man in our representative sample found out about his current job from a work-related or community-based female contact. And only one man reported finding his current job through a female friend or acquaintance. A similar, although less extreme, pattern obtains for women: the bulk of women's nonfamily contacts are other women. Because job information flows through gender-biased networks, the jobs women are most likely to learn about are the jobs that their female informants know about or have held, and, given women's general

Table 7.3 Types of personal contact used in obtaining present job

			Women in occupations		
	Women	*Men*	*Female-dominated*	*Gender-integrated*	*Male-dominated*
Sex of contact					
Female	49	9	57	45	29
Male	34	70	28	36	52
Unknown	27	25	29	25	29
Origin of contact					
Family	36	26	36	33	43
Community	24	6	29	17	19
Work	22	30	18	27	24
Friends/acquaintances not specified as either work or community related	29	42	30	27	24

Source: Hanson and Pratt 1991; personal interviews, Worcester, MSA, 1987
Note: Figures are percentage of each group having used each type of contact, e.g., 49 percent of women who had received job information from a personal contact obtained that information from another woman. People could report having used more than one contact

propensity to work close to home, these job opportunities are most likely also close to home.

Gender-biased networks are also likely to perpetuate sex-based occupational segregation, as women inform other women about female-dominated jobs and men tell other men about male-dominated ones. As Table 7.3 shows, women in female-dominated occupations are more likely than other employed women to have gotten job information from other women, and they are the most likely to have received that information from someone in their local community. Among the employees we interviewed at firms in 1989, there was a very strong tendency for these personal contacts to draw others into the same occupations: 29 percent of the employees, and 37 percent of the women, we interviewed with friends or relatives at work had the same three-digit census occupation code as their friends or relatives.

Given their relatively unique status, it is especially interesting to note that women in male-dominated occupations have relied very heavily on personal contacts, but ones that differ noticeably from those drawn upon by other women. Three out of four women in male-dominated occupations had used a personal contact to obtain their jobs; this compares to half the women in other occupations. Further, women in male-dominated occupations were much more likely to have received help from men, almost half (43 percent) of whom were family members. This generalization may be more complex, however, than it first appears, with women playing a more active role in some of these cases. One of our interviewers happened to interview his aunt and, upon reflection, qualifies a response that was no doubt coded as a male contact:

> When I did my aunt's interview I found out things I didn't know. Like my father got her a [production] job at Nortons. But I'm sure that it was really my mother [who pressured my father to get the job for my aunt]. She was her sister and she was looking out for her. My aunt was the youngest and didn't have kids and was knocking around the house.

Regardless of this qualification, the fact that women so often depend on a male family member for entry into a male-dominated occupation calls into question the generalizability of the conclusions drawn about the role of networks in job search in earlier studies based on all-male samples (e.g., Lin and Dumin 1986; Granovetter 1974). These studies have concluded that contacts with friends and acquaintances or people from work more often lead to jobs with upward occupational mobility than do familial contacts. We have found that, for women, relatives can play an important role in opening up opportunities in gender-atypical occupations. For example, one women in our sample, who worked as a salesperson selling welding equipment for a welding supply company,

Table 7.4 How people found their jobs

| | Occupational standing | | | | | | | |
| | Non-skilled | | Skilled | | | | | |
Ways of finding job:	manual (%)	nonmanual (%)	manual (%)	nonmanual (%)	Managerial/professional (%)	p-value
Impersonal						
Women	35.5	36.5	35.0	43.4	43.3	n.s.
Men	24.3	15.4	23.9	30.4	26.5	n.s.
Contact with no authority to hire						
Women	58.1	43.5	50.0	30.0	21.7	.002
Men	48.6	69.2	32.6	17.4	26.5	.001
Contact with authority to hire						
Women	12.9	14.8	5.0	15.9	23.3	n.s.
Men	16.2	15.4	23.9	13.0	26.5	n.s.
Union/employment agency						
Women	0.0	0.9	5.0	12.4	20.0	.000
Men	24.3	0.0	15.2	15.2	12.9	.001

Type of contact:

Relative						
Women	29.0	25.2	25.0	13.3	6.2	.007
Men	21.6	23.1	10.9	8.5	8.2	n.s.
Community						
Women	22.6	16.5	5.0	13.3	10.0	n.s.
Men	10.8	15.4	8.7	4.3	4.1	n.s.
Work						
Women	3.2	5.2	15.0	18.6	18.3	.002
Men	8.1	23.1	10.9	21.3	20.4	n.s.
Unknown						
Women	22.6	15.7	15.0	8.8	3.3	.03
Men	27.0	30.8	17.4	2.1	10.4	.006

Source: Personal interviews, Worcester MSA, 1987

Note: Occupational standing refers to individual's own occupational standing. The *p*-value is from a test of differences across occupational classes. Percentages indicate the proportion of employed women or men who found job this way. People could report more than one way of finding a job. n.s. = not significant

had been hired by her father, who owned the company: "He gave me the job." Husbands are often the male family members who draw women into male-dominated occupations, as in the case of a 47-year-old woman who had worked for thirteen years repairing valves at a large valve manufacturer: "My husband worked there; that's why I'm working there. I needed a job, and I kept bugging personnel until they hired me. They like to hire family there – lots of husbands and wives, fathers and sons." The experiences of these women are, however, exceptional: fewer than 10 percent of the employed women in our sample were working in male-dominated occupations. These women's experiences also underline the continuing strength of patriarchal relations both at home and work, and the boundaries that restrict women's access to male-dominated occupations. In Worcester, these boundaries are often permeable only through the consent, help, or reputations of male family members.[16]

As a third generalization, strategies of job search vary for both men and women in different occupational grades. The tendency to find a job through a contact who has no authority to hire is much more common among men and women in lower occupational grades (Table 7.4).[17] On the other hand, job search through a professional organization (even including unions) is a strategy employed almost exclusively by men and women in professional and managerial occupations. The geographies of these two strategies of job search are likely very different, with the first more closely tied to the local community, where friends, relatives, and neighbors might trade information about job openings. This impression is at least partly confirmed when one looks at the characteristics of personal contacts drawn upon by persons in different occupations (Table 7.4). Relatives are a much more common source of job information among women in nonskilled and skilled manual and nonskilled nonmanual occupations; the same trend exists for men, although it is not statistically significant. So too, although the difference is not statistically significant, the trend for both men and women is for those in nonskilled occupations to rely much more heavily on community-based contacts for job information.[18]

The finding that both men and women in lower occupational grades are more likely to find their jobs through relatives and community contacts resonates with the earlier discussion, in Chapter 5, in which we document the tendency for men and women in lower occupational grades, and particularly manual jobs, to have lived more of their lives in Worcester and to have had their housing passed on to them through relatives. There is a distinctive geography to these class practices: while only 4 percent of the households in typically nonmanual, middle-class Westborough obtained their house from a family member, this was the case for fully 19 percent in working-class Main South and 16 percent in

202

the households interviewed in the Blackstone Valley. It is clear that many households living in working-class communities of Worcester are embedded in long-term networks of very local social relations that shape many aspects of their lives, certainly their manner of finding housing and jobs. The networks that not only lead people to housing and jobs but also affect "just everything" (as Michael Brown put it, see remarks in Chapter 3) reinforce the localism that characterizes the lives of many Worcester residents and are part of the processes that create distinctive communities of employment in different parts of Worcester.

As a final observation, the generalizations that we have made about the geographies of job search are themselves conditioned by the specificity of particular places. In general, those who search informally, especially women and workers in lower occupational grades, find jobs close to home. Recognizing the localized nature of the majority of individuals' job searches helps us to understand how local labor markets develop and how a patchwork of distinctive social and employment worlds develop within a single metropolitan area. Nevertheless, these generalizations about the geography of job search, drawn in terms of the social categories of gender and class, are mediated by characteristics of specific places. A close look at households living in the Blackstone Valley helps make this point clearer.

Residents of the Blackstone Valley tend to be rooted in the place,[19] and many women (41.4 percent) and men (33.3 percent) had found their job through a friend, relative, or neighbor. As a point of comparison, 21.4 percent of men in the nearby middle-class suburb of Westborough had found their jobs through a friend, relative, or neighbor. Nevertheless, men living in the Blackstone Valley travel long distances and times to work, slightly farther and much longer than men from Westborough (15.4 miles and 15.0 miles, and 34.5 minutes and 25.4 minutes, respectively). The characteristics of the job locations of men living in the Blackstone Valley deviate from what we would expect because of the nature of the place. Living in "the cradle of deindustrialization" (Reynolds 1991: 177), many residents of the Blackstone Valley have been forced to leave the Valley for employment. As we noted in Chapter 3, two-thirds of employed local residents worked outside the Valley in the mid-1980s. The commute, in these cases, is lengthened by the geography of transportation: traffic between the Blackstone Valley communities flows mostly along Route 146, which runs between Worcester and Providence, Rhode Island along the Blackstone River but is unconnected to the Massachusetts Turnpike, the state's major east–west route into Boston (Figure 2.1, p. 30). Blackstone Valley men tend to compensate with long commutes for these accidents of geography and for being tied through social relations to a place that seems to offer ever fewer employment opportunities. The characteristics of actual places, therefore, sometimes disrupt the geographies of distance that we have described. As a generalization, however, the use of local networks for job information – so prevalent among those living in Worcester – sustains local and spatially disjointed labor markets and sex-based occupational segregation.

LOCALLY AVAILABLE JOBS AND GENDER, RACE, AND CLASS RELATIONS

We have discussed in Chapter 4 how locally available employment opportunities affect occupational choices for some groups of women, in particular women with small children who are working part time. We want to extend this discussion to consider the many ways that employment opportunities vary across local communities – even within female-dominated occupations – in ways that affect women's employment patterns, gender relations, and family life.

Community and occupation

Different areas of Worcester offer very different employment opportunities. Because women tend to find jobs so close to home, this variability in locally available jobs has a nontrivial impact on the types of work that women do in different areas. Women who work in manufacturing and producer services firms in the four case study areas, for example, tend to do very different jobs, most of which are nevertheless in female-dominated occupations (Table 7.5). Three out of four women employed in Upper Burncoat firms, and one out of two in Main South and the Blackstone Valley, are employed in production jobs, while fewer than one out of every four women in Westborough firms do this kind of work.

Some of these differences, particularly the tendency for so many Westborough women (almost one in four) to find employment in managerial and engineering occupations, no doubt reflect differences in educational and class backgrounds.[20] But the class characteristics of many occupations are not given or easily attributed to specific educational qualifications; they are constructed in place. They are constructed in place as employers locate their businesses so as to seek out women with particular social characteristics (e.g., middle class, white) to do some types of jobs (e.g., clerical) and those with other qualities (e.g., working class, mill culture) to do others (e.g., assembly). They are constructed in place in part because of the relative spatial immobility of different groups of women.

The case history of one woman living in the Blackstone Valley conveys a sense of how locally available employment opportunities direct women into particular types of female-dominated occupations. This woman,

Table 7.5 Types of work that women do in the firms surveyed in four Worcester areas

	Westborough (n = 42)	Main South (n = 10)	Upper Burncoat (n = 16)	Blackstone Valley (n = 31)
Total number of jobs in the firms surveyed	2,589	1,504	1,251	989
% jobs filled by women	27.7	32.0	45.9	40.7
% of all employed women working in:				
General unskilled (e.g., cleaner)	6.4	9.3	0.8	8.2
Skilled and unskilled production	20.5	54.3	73.5	46.7
Engineering	10.2	0.0	0.0	2.2
Clerical and sales	50.1	29.1	20.0	35.5
Managerial	12.8	7.2	5.6	7.4

Source: Interviews with Worcester-area employers, 1989

now 50 years old, immigrated to the Blackstone Valley in the late-1960s from England, along with her husband. She had done office work in England, and when her husband was laid off from his job in 1971, this was the type of work that she sought. She preferred certain types of office work over others; in particular, she found typing boring and wanted to work as a receptionist. As she had young children and no car, she needed a job close to home with "mother's hours." Her neighbor told her of a clerical job at her workplace; when she obtained this job, she walked to work with this female neighbor. When computers were brought into her workplace, she "couldn't get it" – we have no details about the training that was made available to her – and she was moved into production work as an assembler. The move was clearly a demotion, involving a shift from salary to hourly wage, and reductions in pay and benefits. This woman preferred office work and returned again and again, throughout the interview, to her loss of status: "Now I'm on the same level as everyone else at the plant today. I'm just assembly. If a machine is down, I do another [assembly] job." She described her current job as "a bore. . . . It's a pay check."

This is not simply a story of a woman who was unable, for whatever reasons, to upgrade her skills to maintain her office job. When she got a car, she attempted to remedy her situation by looking around for office work: "I looked around a little, but the majority of things that interested me . . . they were mostly around Worcester." Based in the Blackstone Valley, she judged these jobs to be too far away (though, in fact, downtown Worcester is less than ten miles from her home) and elected to stay with her job as assembler. One can only speculate, but it seems likely that if this woman were living in Worcester, she would not be working as an assembler.

The availability of clerical jobs may depend on the actual composition of jobs in a local area, but the reputation of the area can also affect individuals' desire to take up available jobs. In Upper Burncoat, potentially permanent clerical jobs remained unfilled because of the area's reputation. An employer in this area spoke of the difficulty of finding clerical workers because of the firm's proximity to Great Brook Valley, the stigmatized public housing project. "We have found that it's difficult to get office help. Being next to Great Brook Valley has hurt us. Some people won't even show up for interviews, some people are just too afraid." Unable to find permanent clerical workers, he had resorted to the use of temporary workers to fill clerical positions. (We did not pursue the issue of whether assumptions on the part of the employer had led him to exclude local women living in Great Brook Valley as potential clerical employees. In general, he was not adverse to hiring residents of Great Brook Valley as production workers, because almost all of his production workers, including several women, came from this housing project.) What this example demonstrates is the power of place – both in fact and reputation – to structure employment opportunities.

Many women with whom we spoke in areas where production jobs predominated preferred to work on the factory floor, and their struggles took place around entry into male-dominated occupations. In the Blackstone Valley, for example, prized male-dominated occupations are typically production ones. A 46-year-old woman, working as a blanket cutter in the Blackstone Valley, told us how she had been informally trained (and exploited) as a cutter at her previous job in the Blackstone Valley. "They told me to train a guy, so I was training him. I was doing all that plus my own job. Then someone told me that I was training him to be my boss . . . I said I worked a man's job. Then I went to an agency to find out what other cutters were getting. They told me $7 to $12 an hour. I was only making $5 an hour, so I left that job after they would not give me more money." She currently works as a cutter, the highest-paid production job in her factory. "I kind of learned on my own. They always had men doing this job, but I told them I could do it."

The gendering of jobs is clearly under contest at her current firm. Five years ago, cutting in the blanket department was a male preserve; now seven of the nine cutters in the department are women: "We women are taking over . . . Two women started in the shop before me. . . . They [the men] weren't too happy having women come in. They didn't treat us too well."

Whether this woman has finally won her continued struggle for recognition of and compensation for her talents and hard work is an open question. The regendering of the cutting job in this particular factory has followed changes in the technology of cutting, which make it a "less strenuous, less acrobatic" job. The cutter's job in this factory may follow one of the trajectories that Reskin and Roos (1990) note in their study of women's entry into male-dominated occupations: male-dominated occupations to which women gain entry are often quickly feminized, and the wages then stagnate. Despite the persistence of such discouraging outcomes, we should not lose sight of the efforts taken by individual women to gain access to male-dominated occupations: in this case, researching wages for cutters, confronting her employer with evidence of unfair wages, quitting her job because her employer was unwilling to compensate her fairly, and enduring sexual harassment from male co-workers.

Of particular interest to us is the fact that different contexts allow women different opportunities for entry into male-dominated occupations. In their study of women's entry into male-dominated work, Reskin and Roos (1990) (and other contributors to their volume) display some sensitivity to context. In Steiger and Reskin's case study of the occupation of baker, for example, they notice that the feminization of the occupation has occurred less often in traditional bakeries and more often in modern bake-off bakeries in supermarkets and retail chains. There are further questions of context and geography to be explored, however, in particular, the uneven geography of women's access to male-dominated occupations.

Certainly, the firms in our four case study areas have allowed women different degrees of access to traditionally "male" occupations. In Westborough, 22.1 percent of all jobs in occupations classified as male-dominated in the firms surveyed were filled by women. This was the case for 16.5 percent and 15.2 percent of jobs in male-dominated occupations in Upper Burncoat and Blackstone Valley firms, respectively. In marked contrast, in firms in Main South women fill only 9.2 percent of jobs classified as male-dominated. What this suggests is that it is more difficult for women living in Main South to gain access to male-dominated occupations because the only ones available locally are inaccessible to them. This may result from the types of male-dominated occupations most prevalent in the area, occupations such as machinist and mechanic. But to argue in this way begs the question of why there is such resistance to recoding the occupation of machinist, for example, in gender-neutral terms.

A number of factors seem to be at play. First, some employers in Main South were clearly operating with sexist stereotypes. From an employer of a sheet metal company, we heard: "We are supposed to have 5 to 6 percent females by federal guidelines but we have only 2 percent. Our work is very physical. If there's a woman on the job, who gets the bathtub on the second floor?" Second, a number of Main South employers mentioned their efforts to retain their skilled labor force, the stability of their present employees, the reduced size of their labor force, and the fact that their labor force was now middle-aged and ageing. In circumstances of low labor turnover and infrequent hiring, there is little opportunity for gender recomposition of occupations, at least in the short term. Third, when hiring did take place, the use of personal networks of the current workforce for labor recruitment was extremely common; this would tend to reproduce the existing gender bias. Fourth, a number of employers who were hiring and who had experienced labor shortages in the mid-1980s had identified a newly arrived Vietnamese labor pool (resident within Main South) as ideal, and employers seeking low-wage labor had adopted the strategy of hiring men from this particular ethnic network. Finally, many of the firms in Main South are old ones that have operated in the area for generations; the weight of tradition may work against changing the gender composition of traditionally male jobs. A number of these factors – the availability of a new immigrant labor pool within the neighborhood and cultures of expectation among neighborhood employers – are structured within the local context. It is possible that an occupation such as machinist is more permeable to women in places other than Main South.

We do find that women have access to different types of traditionally male occupations in different areas, suggesting that processes of labor market segmentation are not only geographically variable, as both Peck (1989) and Morrison (1990) speculate, but that this variability occurs at a very fine geographical scale. In Westborough, a substantial proportion (although by no means equal to men's) of managerial and computer systems analyst[21] jobs are filled by women (21.5 percent and 21.3 percent, respectively). In the other areas, women are rarely managers, although they have broken into other types of traditional male employment (Table 7.6).[22] In Main South, almost one out of every five machine operators working in metals and plastics is a woman. In Upper Burncoat, women have been drawn into technical occupations from which they are usually excluded. One-quarter of all science technicians and draftspersons, 41 percent of manufacturing sales representatives (for products such as valves), and almost one-third (29 percent) of all production supervisors are women. The long-term implications of women's presence in particular male-dominated occupations for relations between men and women in each area is unclear, especially given Reskin and Roos's (1990) pessimism concerning long-term gender integration within particular occupations. What seems clearer is that the spatial patterns of permeability reinforce differences among women,

Table 7.6 Women in male-dominated occupations in the four study areas

	Westborough women (n)	Westborough jobs (n)	Main South women (n)	Main South jobs (n)	Upper Burncoat women (n)	Upper Burncoat jobs (n)	Blackstone Valley women (n)	Blackstone Valley jobs (n)
Managers	26	121	19	100	3	40	11	85
Computer technician or analyst	38	178	n.a.	0	n.a.	0	n.a.	0
Science technician	3	35	2	10	29	114	3	12
Sales – insurance	6	29	n.a.	0	0	9	n.a.	0
Sales – manufacturing	0	13	3	16	31	76	7	36
Machine operators (metal, plastics, lathe)	12	25	21	117	n.a.	0	n.a.	0

Source: Interviews with Worcester-area employers, 1989
Note: Managers include 1980 Census Occupation Codes 06, 013, 019, 026; science technician includes 1980 Census Occupation Codes 189, 213, 217, 224, 235; machine operators include 1980 Census Occupation Codes 703, 704, 715, 725, 794. Number of women and number of jobs refer to total numbers for each occupation for the interviewed firms in each area. n.a. indicates not applicable

with women in middle-class, nonmanual areas expanding their access to certain types of "managerial" and skilled nonmanual jobs, and women in working-class areas filling a broader range of production occupations. Patterns of access to traditionally male occupations then reproduce and extend existing area- and class-based differences between women.

Race, class, and discourses of gender

Employers in different areas also seem to hold different stereotypes about special capabilities that women may have as waged workers; these stereotypes may stand in for and articulate other preconceptions about race and class. We can thus see gender acting as a type of "transfer point" for other social relations.

Some evidence for this assertion comes from employers' responses to our queries about whether women have certain skills or other character-istics that they value in workers (Table 7.7). In Westborough, many employers were simply unwilling to reflect on the special characteristics of women as workers. This, in itself, is interesting and may reflect the fact that we more often spoke with human relations officers in Westborough

Table 7.7 Percentage of employers citing trait as skill or characteristic that they value in women

	Main South (n = 37)	Upper Burncoat (n = 10)	Blackstone Valley (n = 24)	Westborough (n = 33)
None	22	40	17	46
"Nimble fingers"	22	20	8	9
Willing to do repetitive work	27	10	4	12
Docile	19	50	12	15
Hard workers	3	10	21	21
Loyal	24	0	29	21
Cheaper	8	10	8	3
More careful	11	10	29	12
Can't push too hard (negative characteristic)	5	0	33	16

Source: Pratt and Hanson, 1994
Note: These answers were given in response to question: "Do women have certain skills or other characteristics that you value in a worker?" We have classified the variety of responses into the categories shown in this table

than in other areas (as opposed to managers and owners), and they may have been more savvy to the politics of image management. But when Westborough employers did indicate that they value women employees for certain qualities, they emphasized personal ones: they mentioned that women are hard working, dependable, loyal. These are rather positive, respectful, perhaps "wifely" characteristics. In Main South, women are also valued for their stability and loyalty but employers also mentioned women's capacity to doing boring, repetitive work and their superior manual dexterity. Upper Burncoat employers speak of women workers' "nimble fingers" but even more stress their pliability and docility. Blackstone Valley employers most often mention women's capacity to do careful, detailed work, as well as their stability and loyalty. Many spoke of women's physical weakness, telling us that: "Women can't do heavy work" or "You can't push women as hard." In the last three areas there is a physicality to the stereotyping, undoubtedly reflecting the higher proportion of manual work done by women in these areas. In Main South and Upper Burncoat, stereotypes of "nimble fingers," tolerance for tedium, and docility no doubt operate at the intersection of race, class, and gender, reproducing pervasive clichés about the skills of Asian workers (see also Scott 1992a). This very partial evidence suggests that women are conceived as employees somewhat differently in different areas, in part because of their coding in terms of race and class.

Employment context and household relations

For many individuals, the local employment context affects not only the occupations that they "fall into" but their options for balancing work with others in their household. Studies of household–labor market relations have focused on the ways that national (Morris 1990) and regional (e.g., Morris 1990; Wheelock 1990) contexts shape the organization of work among household members, but very little attention had been given to how these processes might work at a finer scale such as communities within urban areas. We draw attention to three characteristics of employment that vary from place to place within the metropolitan area: schedules of paid employment, the availability of part-time work, and wage levels. We consider how variations in employment conditions affect how households structure their lives and the impact that this has on gender relations.

The timing of shifts and schedules are variable across places, such that particular household work strategies are more or less possible in different communities within Worcester. Shift work is much more readily available, to women and men, in Upper Burncoat and the Blackstone Valley (Table 7.8). The availability of evening or night work to women is tied not only to the gender characteristics of jobs that get done at those times

Table 7.8 Opportunities for part-time and shift work in manufacturing and producer services firms in four local areas in Worcester

	Westborough (n = 42)	Main South (n = 40)	Upper Burncoat (n = 16)	Blackstone Valley (n = 31)
% firms offering part-time work	75	57	56	80
% firms hiring women part time	70	45	50	48
% firms that have evening or night shift work	26	28	50	52
% firms that have women working evening or night shift	11	7	38	26

Source: Pratt and Hanson 1994; interviews with Worcester-area employers, 1989

211

but also to the specifics of place. One of the reasons that so few Main South employers hired women to work on the evening or night shifts is intrinsically related to the characteristics of the place: it has a reputation as a dangerous part of town. Two-thirds (62 percent) of Main South employers who run night shifts said that they do not hire women to fill them because women have no desire to work the night shift, in part because of the location.[23] In the words of one Main South employer: "People are uncomfortable leaving here at four o'clock [in the winter] when it's dark." From another: "There are no women on the night shift because they are afraid. Would you want to come down here at night?" The reticence of women who work in Main South to travel to employment at night is no doubt compounded by the fact that more than one-quarter (26 percent) of the women interviewed in Main South had no access to automobiles and therefore had to travel through the area by bus or on foot.[24] Thus while employers in the Blackstone Valley are known to advertise: "Work while your kids are asleep, the Mother's Shift," Main South firms that advertise "Mother's Hours" refer to a shift that runs from 9:00 a.m. in the morning to 2:00 or 3:00 p.m. in the afternoon. Also in contrast to Main South, an employer in the Upper Burncoat area (located in what he described as a campus-like setting, some distance from Great Brook Valley) noted that one of the attractions of the location was that employees could work there at night. This insurance company had moved from New York City to Worcester in the late 1960s and the interviewee observed that: "[In New York] the loss of productivity is tremendous. In New York, they don't work late at night for security reasons. It's safer in Worcester, better lighting. There's paid security around the parking lot. Here, we have fifty cars in that lot at night." The contrast could easily have been made closer to home: many employees in Main South are no less worried about night-time security.

These differences in the scheduling of employment have implications for access to certain types of occupations: a number of employees interviewed in Main South mentioned that the second and third shifts are the entry-level shifts for some of the male-dominated occupations. If women do not have access to these shifts because of their (or their employers') concerns about safety, they are effectively denied access to these male-dominated occupations.

Area-based differences in work schedules also affect the ways that households arrange work. It is notable that only employers in Upper Burncoat and the Blackstone Valley mentioned that women and some-times men work evening or night shifts so that child care can be shared between partners. In these areas, from one-fifth to one-quarter of all employers offering shift work spontaneously mentioned the sharing of child care within employees' households. The following are some of the comments made by employers in the Blackstone Valley about the accommodations that their workers make around child care and shift work. From an employer at a mill that manufactures pile cloth: "Women work the second and third shift. They ask for it. It's good scheduling to watch children. They either get someone, or the husband watches the children while they work." From a manufacturer of yarn: "[The women who work from 6:00 p.m. to 10:00 p.m.] have young kids and their husbands are home during these hours to watch them." From the owner of a spinning mill that hires women for the second shift: "A lot of people like the second shift. It's because of the way home life is structured." And finally, another Blackstone employer notes: "Some guys would rather work nights with day care concerns. Being flexible as you can with that stuff, the petty-ass stuff is put to rest." We have no illusions that gender relations between men and women are any more equitable in Upper Burncoat and the Blackstone Valley; our point is simply that women and men arrange domestic work and paid employment in different ways in the four areas and that local employment opportunities at least partially dictate the limits of these arrangements.

There is a further way that geographies of scheduling structure gender and class relations. The availability of part-time employment also varies by area, with more firms in middle-class Westborough hiring women on a part-time basis (Table 7.8). About one-third (29 percent) of firms in Westborough that offer part-time employment said that they do this because women want this schedule. This was also a common response among employers in the Blackstone Valley who offer part-time work (28 percent said that they offered part-time hours because this was desired by women – but note that a much smaller percentage of employers in the Blackstone Valley compared to Westborough offered part-time shifts) but was mentioned only infrequently by employers who scheduled part-time employment in Main South and Upper Burncoat (by 16 percent and 11 percent respectively).

Local community child care resources may partially account for the lack of concern on the part of Upper Burncoat employers to offer part-time employment to women. Full-time subsidized child care is available to many women living in Great Brook Valley, which allows them the opportunity to work full time. As we noted in Chapter 6, several employers acknowledged how they benefited from this service, recognizing that it made available to them a relatively low-cost, full-time female labor force.

The geography of part-time scheduling also complements our earlier observation (in Chapter 5) that it is typically women married to men in higher-status jobs who are employed on a part-time, rather than full-time, basis. There is no doubt a synergistic relationship between class and local availability of part-time employment: because of their partner's

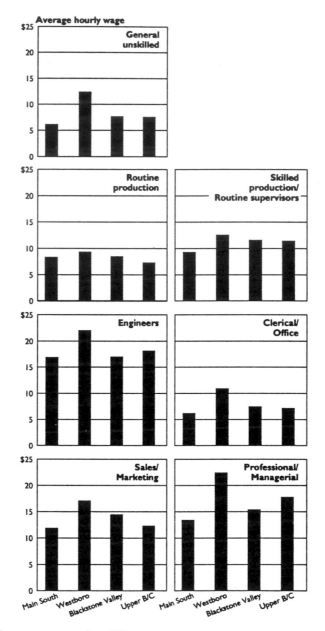

Figure 7.1 Average wages for different types of work in the four study areas: general unskilled; routine production; skilled production/routine supervisors; engineers; clerical/office; sales/marketing; professional/managerial. For tests of significance in differences in wages among the four areas see note 25

Source: Personal interviews with employers in manufacturing and producer services firms, Worcester, 1989

relatively high incomes, middle-class women can afford to work outside of the home part time; local firms in Westborough offer part-time hours to attract female workers; part-time work is then more readily available to middle-class women. Local employment opportunities then support and enable a particular middle-class ideal of arranging mothering and paid work, one in which paid employment is fit within children's hours of schooling, and there is minimal disruption to full-time mothering work. Our point is that this ideal is more easily lived within Westborough than in the other areas, not only because more Westborough households can more readily afford it, but also because more Westborough employers offer this type of schedule. Employment arrangements differ from area to area, and these differences both reflect and nurture different class-identified cultural practices and gender relations.

The extent to which areas shape material resources within households, and hence work and gender arrangements, becomes exceedingly clear when wages are compared across areas within our 1989 sample of manufacturing and producer services firms (Figure 7.1). Employers reported the average wage for each detailed job category in their firm. We then aggregated these jobs into seven general job types and calculated the average hourly wage for each job type in each category. Differences across areas in average wages for the same job types are quite startling, with firms in Westborough paying the highest wages, and those in Main South paying the lowest.[25] For example, a person (most likely a woman) employed in clerical/office work in Main South earned, on average, $6.21 per hour. A person doing comparable work in Westborough earned almost twice this amount ($10.93 per hour).[26]

The implications of these wage differences are quite interesting – and far-reaching. Because the large wage differentials between middle-class Westborough and the two working-class areas exist across all job types, it is clear that both men's and women's wages are lower in the working-class areas *for doing comparable work*. This means that a man doing, for example, general skilled work in Westborough may bring home a wage that allows his wife the possibility of working outside the home on a part-time basis only. (Then, too, a woman working part time in Westborough can earn almost as much as one who works full time in Main South.) Men working in the same job types in other areas are unlikely to earn enough to allow this possibility. That is, the class effects of Westborough go beyond the simple availability of more nonmanual, middle-class jobs: persons doing almost any job in Westborough are more likely to be reproduced as middle class (in terms of cultural practices) because the wages are higher. This then has implications for women's paid employment (for example, whether it is full time or part time) and for the distribution of work within households.

215

Community resources and maternity-related breaks

In the academic literature that assesses the impact of child-related breaks on women's career trajectories, it is typical to conceive of the break from the labor force as an experiential void, during which time the mother withdraws from any meaningful public life that might bear on later job skills or preferences. (See Chapter 5 for a brief discussion of this literature.) Indeed, through our silences, we are guilty of this underlying assumption in our first look at the impact of child-related breaks on women's employment in Chapter 5, in which we focus on the relationship between breaks and shorter home-to-employment commutes. This assumption is problematic, however, because it reproduces the pernicious dichotomy between public and private life that so many feminists have criticized and is as well a very narrow, economistic reading of social life. Once we acknowledge the richness of social life, both inside and outside of labor market experiences, our attention is drawn to the content of experiences during child-related breaks and to how neighborhood context mediates those experiences. We turn now to refine our understanding of the impacts of child-related breaks on women's employment careers by considering how neighborhood context shapes the experiences of having a child-related break, as well as the effects of a shorter home-to-employment commute that often follows a child-related break.

Consider the experiences of a woman living in a white-collar suburb of Worcester. This woman had minimal job experience before taking a break from the labor force. She had held one job for a two-year period: at age 24 she began working as a secretary/editorial assistant[27] for a university press in Cambridge, Massachusetts. When she married at age 26, she quit her job. She had the first of two children two years later and chose to remain unemployed for a further six years, returning to formal employment around the time that her second child was born. She could hardly be said to have been idle during this time, however, engaged as she was not only in childrearing but also in a variety of volunteer activities. Her current job as a writer and reporter for a local newspaper emerged out of one of these volunteer activities:

> I didn't find them [her current employers]. They found me. I'd been attending meetings of the [town] Board of Selectmen as a member of the Worcester League of Women Voters. A study done by the League of Women Voters was the catalyst for the creation of our [news] paper. My knowledge of selectmen affairs made me a natural to write about that aspect of town affairs. I had no journalistic background at all. The woman who became the editor of the [newspaper] was also a League member. It was by her request that I came on board. The League of Women Voters was doing a study of

communications between residents and town officials. . . . The League of Women Voters takes up all kinds of issues. I also started taking up issues about education and those kinds of things, things that were related to my children's well being.

In the future, this woman plans to write "more serious investigative stuff; less human interest stuff, or maybe [do] more public relations." There are two aspects of this personal history that are interesting: first, during her so-called "child-related break" this woman was developing job skills and contacts outside of the labor market or formal educational institutions that were decisive for her later occupational mobility. Second, her choice of volunteer activities was both class and neighborhood dependent. Women in working-class neighborhoods also develop rich social networks during child-related breaks, but their contacts may not have the same influence or their neighborhoods the same resources to propel upward career mobility.

To explore this idea more systematically, we mapped histories of career mobility throughout the metropolitan area for women who were interviewed in 1987 and who had returned to employment from a child-related break. As a surrogate for neighborhood social class, we used 1980 census data for household income, and divided lower-income from higher-income census tracts.[28] This is no doubt an overly stark division, given that there is income variability within every census tract. Nevertheless, women who were living in different parts of the metropolitan area tend to have very different histories of mobility after a child-related break (Table 7.9). We measured occupational mobility by comparing the TSEI (the interval-scale occupational prestige score) associated with a woman's pre-break job with that of her post-break job; only those with an increase (or decrease) of at least ten TSEI points were considered as experiencing occupational mobility. As examples, among those with upward mobility were women who went from being a cashier to a nursery school teacher, from accountant to the chief assessor for a large suburban town, and from secretary to manager of a rest home. Among those experiencing downward mobility were women who went from being a long-distance telephone operator to a cleaning person, from licensed practical nurse to in-home child-care provider, and from officer manager to data entry clerk. Although by far the greatest numbers of women, regardless of residential location, were classified as having no occupational mobility over their child-related breaks from the labor force, women living in higher-income census tracts at the time of their breaks were more likely than those living in lower-income areas to experience either upward or downward mobility (Table 7.9).

That women living in higher-income neighborhoods are at greater risk than those in lower-income neighborhoods for both upward *and*

Table 7.9 Occupational mobility over a child-related break from the labor force, for women in different residential settings

Residential location	Upward (n = 17)	Downward (n = 16)	None (n = 96)	Totals
High-income census tract	11 (17.7)	11 (17.7)	40 (64.5)	62
Low-income census tract	6 (8.9)	5 (7.5)	56 (83.5)	67

Source: Personal interviews, Worcester MSA, 1987
Note: Figures in parentheses are row percentages. "Occupational mobility" compares the TSEI of a woman's pre-break job with that of her post-break job. Upward mobility includes women with an increase of at least 10; downward, those with a decrease of at least 10; all others are classified as having no occupational mobility.

"Residential location" refers to the median family income of the census tract of residence at the time of the child-related break. High-income tracts are those above the 1980 MSA median ($20,180); low-income tracts are those below. Only women with both pre-break and post-break residential locations in the Worcester MSA are included. $\chi^2 = 6.19$; *d.f.* = 2; *p* < .05

downward occupational mobility over a child-related break is interesting and may reflect the fact that many women in middle-class neighborhoods have jobs of relatively high occupational status before a break and therefore have "more to lose" by being out of the paid work force. In fact, the 16 women who experienced *down*ward mobility had a significantly *higher* average pre-break TSEI score (47.9) than did either those who experienced upward mobility (28.8) or those with no occupational mobility (30.9) (*p* < .01). More relevant to our concern with local context is the positive relationship between living in a higher-income area and upward occupational mobility despite the fact that these middle-class women are also more likely to return to long-term, part-time employment (this is the case for 36.9 percent of women living in high income suburbs; Table 7.10), a scheduling strategy that is typically tied to no or downward mobility.

It is arguable that these patterns partially reflect the fact that middle-class women have the educational and cultural resources, independent of their experiences during child-related breaks or neighborhood resources, that afford them upward occupational mobility. In fact, we find no relationship between formal education and occupational mobility, as measured by a 10-point increase in TSEI score from pre- to post-break job (*p* = .56). A woman with at least one college degree is no more likely than a woman with a high school diploma (or less) to experience upward mobility after a child-related break. Note in the previous example that a woman with a B.A. in history, and virtually no relevant job experience, living within the context of a privileged middle-class suburb, developed

Table 7.10 Scheduling of employment after a child-related break

	City of Worcester (n = 76)		High-income suburbs (n = 46)		Low-income suburbs (n = 37)	
Break followed by:						
Full-time employment	35	(46.1%)	21	(45.7%)	15	(40.5%)
Part-time employment	41	(53.9%)	25	(54.3%)	22	(59.5%)
Part-time employment and still working part time	25	(32.8%)	17	(36.9%)	11	(29.7%)

Source: Personal interviews, Worcester MSA, 1987
Note: Figures in parentheses are column percentages. For definition of high-income and low-income suburbs see note to Table 7.9

the contacts and skills through local volunteer activities that later opened the door to a job as a reporter and aspirations to a career as an investigative journalist.

The importance of community resources is even starker for women who may have limited, or dated, formal educational credentials but who have developed long-term roots in a prosperous suburban community. A career in residential real estate is one way that the residential stability and local knowledge of such women can be parlayed into paid work. One 50-year-old woman, who had been a homemaker for 18 years while raising her four children, explicitly recognized a connection between her community knowledge, accrued while she was out of the labor force, and the opportunity to capitalize on that knowledge by working in residential real estate: "I only looked for work here in town. I knew I wanted to sell here in this area – it's an advantage in sales to know the area." Another woman, a 44-year-old now-divorced mother of three, told us how she had started her career in real estate when she was married and her husband had become unemployed. Her first job, which she had held when her older children were very young, had been as a part-time supervisor of roller skating at a local church school, a job that she did "for fun" ("I didn't work there for the money"). But when her husband had lost his job, she needed to earn "a real income." With two years of college and no degree but established roots in her community (ten years in the same house), she had trained as a realtor, obtained her license, and started her own real estate business part time from her house so she could be at home with her youngest son. When she and her husband divorced, she had increased her business to full time.

219

There is a further way that geographical context modifies the impact of a child-related break. We have noted in Chapter 5 that women tend to search for jobs close to home after a child-related break and that a shorter commute (of at least ten minutes) is associated with less upward occupational mobility. When this generalization about the effects of commuting distance is grounded in particular places, it is apparent that it is especially relevant for some groups of women, notably those living in suburban areas. It is suburban women, who typically had especially long pre-child commutes compared to women living in the city, who drastically shorten their home-to-workplace commutes after a child-related break, from an average of 21.3 to 15.2 minutes ($p = 0.00$).[29] A shorter commute within many suburban communities can be particularly restrictive, given the typically lower density of suburban employment opportunities, relative to the city. Among women living in the city, there is no relationship between taking a job closer to home following a break and career mobility. It is particularly within high-income suburbs that a shortened post-break commute is associated with downward occupational mobility: almost one out of three (31.3 percent) women in high-income suburbs who shortened their commutes by ten minutes or more after a break experienced downward occupational mobility, compared to one in 20 (4.3 percent) who did not take a post-break job closer to home.[30] It may be that the low density of both residences and employment in higher-income suburbs sharpens the effects of a shortened commute.

There are two ways, therefore, that geographical context modifies the effects of child-related breaks on women's employment histories. First, neighborhoods provide different resources that women can use to re-enter the labor force after a break. We have argued that the resources available in middle-class neighborhoods are often associated with upward career mobility. Second, given the distribution of employment opportunities in the metropolitan area, women living in higher-income suburbs who choose to shorten their commutes after a child-related break are less likely to experience upward mobility.

CONCLUSIONS

We have argued that the experience of being a woman or man is different, depending on where one lives, because different types of jobs, with different scheduling possibilities, are locally available. Because labor markets are constituted, not only globally, nationally, and regionally but also locally, and because processes of labor market segmentation to some extent get worked out within local areas within cities, women living in different parts of the city have access to different types of employment, including different types of traditionally male-dominated jobs. Beyond

220

employment opportunities, communities offer other resources that shape relations within households and women's access to certain types of paid labor, in different ways in different communities.

It is important to note this geographical variation because it underlines the point that gender divisions and experiences are socially constructed. If material resources and relations allow the women and men of Westborough and the Blackstone Valley different possibilities for living the process of gender, for example, then surely other ways of living gender are possible in different circumstances

8

CROSSING BOUNDARIES

The spatial stories we have told are largely stories of containment. In this final chapter we briefly recapitulate these containment stories and reflect on why we believe it is important to tell them. In the telling, we have ourselves crossed many boundaries – such as those cocooning different literatures, theoretical traditions, and methodologies – and in the third part of the chapter we sketch out how stories of containment have within them the seeds of possibilities for crossing boundaries.

STORIES OF CONTAINMENT

The central plot of our story about the gendered geography of employment shifts from an engagement with space to a focus on place. Like white women in other North American, European, and Australian cities, the women in our predominately white representative sample of Worcester-area residents work closer to home than do men. This generalization masks some important differences among women, even within the composition of this particular sample: working part time and working in a female-dominated occupation both serve to curtail the radius inscribing the geography of women's labor markets. These smaller (than men's) spatial labor markets are etched out on a map of employment opportunities that is itself distinctly gendered: instead of being evenly spread across urban space, women's (and men's) jobs are clustered in particular districts. The unevenness of the employment landscape, together with women's shorter work trips, means that a household's residential location on that landscape importantly defines women's access to jobs. We find that for well-educated, part-time employed mothers of young children, in particular, living in an area with a high density of female-dominated employment increases women's likelihood of working in a gender-typical occupation. Especially for this group of women, then, space does seem to have a hand in occupational segregation. The accumulated facts in this spatial story underline the lead role taken by the friction of distance in differentiating the labor market experiences of women and men as well as those of various groups of women.

We find that a household's residential location, more likely chosen with the man's than with the woman's job prospects in mind, not only has a bigger impact on a woman's (than a man's) access to employment; it also can exacerbate differences among women. Because different residential neighborhoods offer different nonlabor market resources (e.g., volunteer activities, social networks, lifelong learning opportunities), because these resources do have labor market implications, and because a woman's time out of the paid work force is not simply a "black hole" but should be viewed as part of her continuous "work history," residential location – neighborhood context – can affect employment outcomes. In particular, we find that the neighborhood in which a woman lives during a child-bearing/rearing break from paid employment affects the direction of her occupational mobility when she returns to the labor force (i.e., whether her post-break job is "better" or "worse" than her pre-break job). Residential location, then, established more often with men's than with women's jobs in mind, feeds back to affect women's job prospects and to widen differences among women in several ways.

When we look within the household to explore some possible origins of these geographies, we find – as predicted by human capital theorists – strong links between domestic responsibilities and employment situations. Those who shoulder the bulk of the child care and the other work required to maintain a household tend to work closer to home than do those whose domestic workloads are lighter. Although household labor is done disproportionately by women, we find that domestic workloads also have an impact on the work trip lengths of men. Clearly the spatial labor market containment we observe is socially constructed.

While our findings of a strong relationship between position in the home and position in the labor market is consistent with the propositions of the human capital theorists, this "outcome" is not, in our study, the result of the measured decision-making process that human capital theorists have postulated. We find, for example, no evidence that women take jobs in female-dominated occupations as part of a long-term strategy to maximize their life-time earnings; rather, taking such jobs – and taking them close to home – is usually part of a time-management strategy to meet the simultaneous, and very immediate, daily demands of earning a wage and caring for a family. We also find that female-dominated occupations have few of the characteristics attributed to them by human capital theorists (such as higher starting wages). In a similar vein, we find little support for the notion, embedded in the human capital view, that the individual's labor market preferences, expectations, and aspirations take shape outside the labor market; instead, we find that the immediate local environment is often crucial

223

to shaping the preferences and expectations of both workers and employers.

In probing the role of employers in shaping gendered employment geographies, we find that employers contribute both to the unevenly gendered map of employment opportunities and to the gender differences in journey-to-work times and distances that create labor markets of different sizes for women and men. In literally placing employer practices alongside those of workers and potential workers, we begin to see how labor markets are inextricably intertwined with local social and cultural life. We also begin to see how the friction of distance not only helps to generate and sustain gender divisions of labor but also is central to the constitution of different places. Where the friction of distance is particularly strong, these distinctive places emerge at a very fine spatial scale within the metropolis. Localized flows of information, among workers about jobs and among employers about workers, along with a predilection among both employers and employees for relying on information received through personal contacts, help to carve out distinctive, and often highly localized, labor markets. In these, labor market segmentation is literally mapped on the ground.

Offering different types of employment as well as different wages for the same type of work, these places constitute very different milieux within which women and men create and work out labor market aspirations. For women in particular, each of our four study areas holds out different possibilities for crossing the boundary defining sex-based occupational segregation. In one (Main South) numerous jobs in traditionally male-dominated occupations – while close enough to be spatially accessible – remain largely closed to women; in another (Upper Burncoat) certain male-dominated occupations (e.g., science technician; manufacturing sales representative) are clearly permeable to women, while in a third (Westborough) the boundaries around those same occupations are largely impervious to women. These places, then, created in large part through the gendered friction of distance, themselves help to create different experiences of gender and class, different gender and class identities.

The spatial stories we have told are not simple ones, and we want to emphasize their complexity. Are women disproportionately in low-paying, dead-end jobs because they choose them and because they have less education, fewer "skills," and less work experience, or because employers discriminate against women by assigning them certain jobs while reserving better-paying jobs for men? Our Worcester study suggests that, to different degrees in different circumstances, both of these (and other) factors are operating, but that posing the question in an exclusively "either-or" framework is not likely to yield insights that will lead to improving women's position in the labor market.

What is missing from these polarized accounts, as Granovetter and Tilly (1988) have pointed out, is any recognition of the bargaining processes involved or the place-specific social networks implicated in leading to labor market outcomes. Take women's short work trips as an example. These can reflect (simultaneously) a woman's own preferences; her family's priorities and decisions; the job opportunities her family, friends, neighbors, and coworkers know about; employers' screening of potential employees for proximity to the workplace; and employers' location strategies. Take employer location decisions as another example. We have seen how employers do not simply tap distinctive local "pools" of labor; by designing certain labor processes (such as the scheduling and type of work) around particular types of labor, employers actively shape the communities in which they locate. The complexity of our spatial stories comes into focus as our gaze shifts from space to place.

Similarly, the question of whether or not women are hapless victims of a sexist society seems to be framed in an unduly simplified (and unnecessarily polarizing) way (as it is, for example, by Sommers 1994). Certainly the women we interviewed in Worcester did not see themselves, or present themselves, as victims. They were actively making decisions about how best to sustain themselves and their families. Yet they were making these decisions within the existing (and often very constrained) structures of family, friends, employment opportunities, and place.

Space and scale require equally complex accounts. In an attempt to bring conceptual clarity to labor markets, Morrison (1990) has argued that regional labor markets (as defined by commuting range) and local labor markets (as defined by enterprise-level decision making) be distinguished; in a similar attempt, Peck (1989) has argued that the friction of distance and intraurban divisions are relatively unimportant to our understanding of the geography of labor market segmentation, directing attention instead to processes that create different forms and degrees of segmentation in different urban places. Rather than maintaining the empirical significance of these conceptual divisions and giving priority to one category or spatial scale over another (i.e., employee commuting distances vs. enterprise decision making; friction of distance and intraurban segmentation vs. variations across localities), we have argued that it is by looking at how the various geographies are lived and created simultaneously by individuals and firms, through distance and place, that labor market segmentation is better understood.

Throughout this trajectory from space to place, these spatial stories emphasize containment and boundaries. Gender is certainly constructed at many spatial scales, from that of the body to that of the globe; here, we have privileged those scales at which many local labor market practices are worked out. The boundaries around gender-based occupational segregation have a spatial expression in the differently bounded labor markets of women and men. And the gendered boundaries around divisions of labor in the home, around social networks, and around the spaces within which people conduct their daily lives all conspire to create distinctive, loosely bounded places. While telling narratives of containment, we want to emphasize the possibilities, within these stories, for crossing boundaries. We turn, first, however, to consider the significance of these containment stories before outlining how they can be held in tension with eroding boundaries.

THE IMPORTANCE OF CONTAINMENT STORIES

Telling these containment stories seems important for (at least) three reasons. First, these stories enrich the occupational segregation, labor market segmentation, and economic sociology literatures by showing how labor market segments are created in place and how the personal networks that knit together various spheres of life are at once gendered and embedded in place. Bringing these literatures into relation with each other – and with geography – has been productive because it underlines the constructed nature of occupational segregation and the place-based specificity of these constructions. Our containment stories underline the point that space is not a container of labor market practices but is the medium through which people, rooted in places, and communicating with other people with access to different kinds of resources, create distinctive labor market practices that carve out local labor markets within the metropolis.

Second, by telling these containment stories we hope to offer a way beyond the personalist politics that exist within some feminist theorizing. We are sensitive to the potential problems with containment stories, in particular to the charge that calling attention to the extreme localism of many people's, and especially women's, lives sustains a long history of masculinist thinking that conflates "woman" and local, closed and static existences (G. Rose 1993). We have argued, however, that rootedness is a characteristic of most people's, and not simply women's, lives in Worcester and probably elsewhere as well. We also hope that we have shown that the extreme localism that characterizes many women's lives is a fact of existence within patriarchal and class societies rather than a natural attribute of women.

We have drawn attention, not only to gender differences, but also to differences among women and to the role of geography in constructing these differences. We think that it is important to tell this type of geographical story about differences among women because it forces us to recognize the practical difficulties of building collective feminist

226

politics. Women in different places live different schedules of work, have different work experiences and concerns. Mothers in different parts of the city are familiar with different child-care arrangements and household divisions of labor; they live in different moral cultures and have different beliefs about what are acceptable ways of combining parenting and waged work. Organizing across these differences is no simple task. A first step is to recognize the differences and to understand how they are embedded in locally based work, household, and community relations.

Noting this sounds rather like an empty academic formula. But our own experiences with coalitions that try to build affinities across women suggest the need to embody and contextualize differences among women. At a recent workshop (April 1994) around the theme of "Learning and connecting: women as agents of change in their communities," for example, women from many different communities in Vancouver (community groups, nonprofit organizations, community colleges, and universities) came together to discuss the process of working together across our differences. Much of the day's discussion was spent trying to advance beyond personalist politics (e.g., you are a white, middle-class academic and I don't trust you) to try to appreciate each woman's potential contributions by seeing these very different contributions as emerging from the context of each woman's community of experience and relations. We see our work in Worcester as another attempt to do this, to demonstrate in a more generalized way the place-embedded nature of social life and gender relations. When women's lives are contextualized in this way, personalist politics may be less obtrusive.

A third motivation for telling these containment stories is that, with our emphasis on bounded, embodied social life, we suggest not only the real material difficulties of speaking across difference, but the stickiness of identities, grounded as they are in material conditions. This runs counter, we think in a constructive way, to much contemporary feminist theorizing about identity. In reaction to masculinist conceptions of femininity as stasis, many contemporary feminists have been drawn (as have we) to poststructuralist conceptions of identity, as unstable, fluid, and moving. Within popular culture Madonna and Cindy Sherman enact the notion of moving identity, to the delight and fascination of many feminist cultural theorists (Deutsche 1991; hooks 1993), including geographers (Massey 1991, 1994; G. Rose 1993). Massey (1994) summarizes this politics of identity: "One gender-disturbing message might be – in terms of both identity and space – keep moving!"

While sympathetic to the politics of this theory and practice of identity, we offer our study as a reminder of how difficult it is for many women to enact it. We think of the woman we interviewed at an industrial laundry in Main South. Recently arrived from Puerto Rico, she worked and lived

227

within a Worcester Latino community. With little English and long work hours, her chances of occupational mobility, though very much sought after, seemed slim. The neighborhood in which she lived was highly stigmatized; a number of employers told us that her address alone was enough to disqualify her for a job in their establishment. Geography fixes this woman's identity, because she lives in an extremely localized community and because her home place restricts her employment opportunities. Her life experiences seem distant indeed from those of Cindy Sherman and Madonna; we want to force feminist theorists to remember that.

Mobility stories can also disrupt our containment narrative in constructive ways. The woman working in the industrial laundry in Main South, whose life we present as being bounded in a very tight geographical space, had, after all, moved to Worcester from Puerto Rico. At the time that we spoke with her, the tight boundaries of home and work, actively tended by employers and landlords who equated "Puerto Rican" with the very real deprivation in her neighborhood, seemed layered on each other so as to lock her in place. But we really cannot say what adventures in place and of identity she might construct for herself. Certainly, identities are not totally fixed by place, and we were told by some individuals how they actively rejected the stigmatized label of their neighborhood (because they fully understood the effects of such labels). They would avoid telling prospective employers, for example, of their address on Piedmont Street; they refused that geographical label as part of their identity. Mobility stories help to remind us that identities, while constituted in and by places, also exceed them.

CROSSING BOUNDARIES

What do our spatial stories contribute in the way of concrete recommendations for social change? We cannot offer a detailed plan for action, but we can use our Worcester study to evaluate some of the practical suggestions that have been made to cross the boundaries of occupational segregation. We wish to underscore again the characteristics of our sample from which we draw these conclusions, namely mostly white Worcester residents.

Much of what we describe simply provides further evidence for the recommendations that many nongeographer feminists have made for many years. We find that women who do the bulk of domestic work typically find jobs in female-dominated occupations close to home; this speaks eloquently to the fact that a revolution in the home is a precondition to one in the labor market. We find that many women in gender-atypical occupations move into female-dominated ones if they shift to part-time hours after a child-related break; this supports the call for a restructuring of all employment, to allow more flexible schedules, including part-time hours, in a wide range of occupations. We find that male family members are important gatekeepers for women's entry into gender-atypical jobs; this substantiates the importance of mentoring for breaking the barriers of labor market segmentation.

Our focus on geography draws attention to the many ways that space mediates labor market outcomes: familial responsibilities tie many women to an extremely localized labor market, gender identities and norms are developed in place, and labor market segmentation evolves differently in different places. Nevertheless, our focus on geography does not lead us to recommend simple spatial fixes to remedy sex-based occupational segregation; in fact it enables us to see the limitations of some of the spatial fixes that have been suggested by other academics and policy analysts.

Since the late-1960s a number of social scientists have been concerned about the effects of a spatial mismatch that reduces the accessibility of the urban poor to well-paying jobs (Kain 1968; Holzer 1991). These scholars have recognized that many African Americans, ghettoized in inner-city housing markets by class and race discrimination, have poor access to expanding employment opportunities in suburban areas, especially because existing public transportation tends not to service these areas very effectively. This analysis, which diagnoses inner-city unemployment as essentially a spatial accessibility problem, prompted numerous transportation-based remedial programs in the late-1960s and early-1970s, such as special bus services from the inner city to suburban job sites. Some even recommended providing everyone with a Volkswagen "beetle" ("new Volks for poor folks", Myers 1970: 199) as a more efficient and cost-effective solution than mass transit.[1]

Certainly we encountered individuals in Worcester whose work lives had been adversely affected by lack of accessibility to a car. One woman we spoke with in a town southwest of Worcester had been forced to leave her office job in Worcester when friends, who owned a car and with whom she had commuted with to work, moved their residence. Given the limited range of occupations available close to her home, she now worked at a job packing metal handles in boxes. She disliked her present job (conditions were bad, pay was low, and there were no benefits) and would prefer to do office work. She speculated that if she lived in Worcester she "wouldn't work in this dump" but acknowledged that [spatially restricted] "beggars can't be choosers." For this woman, as well as for the other individuals who told similar stories, access to an automobile would indeed effect the transformation from beggar into chooser.

Still, (and leaving aside the issue of environmental impacts) a Volkswagen in every garage is unlikely to disrupt general patterns of

229

occupational segregation. Despite the fact that more than three-quarters of the women in our Worcester study drove alone to work, occupational segregation was still well entrenched. Our point is that, while improved transportation and spatial access may help some individuals in important ways (and especially lower-income individuals, who generally had poorer access to cars), it is not a panacea; access to a car was not an issue for most women in Worcester.

We also question the effectiveness of high-tech attempts to annihilate space by transcending, for example, the fragmentation of job information that results from the fact that such information flows through place-based social networks. In a report recently released by the Canadian Labor Force Development Board, reliance on networks for job search and job recruitment was tied to the "fragmentation" of the labor market. The Board recommends several measures to counteract this fragmentation, including assembling an extensive national database of employment opportunities and making it accessible to prospective employees in neighborhood supermarkets (*Vancouver Sun* 1994: A1; Hughes 1991 makes a similar policy recommendation as a way of making suburban job opportunities known and available to inner-city residents).

Our research indicates that reliance on personal contacts is not an effect of the fragmentation or segmentation of the labor market; it is a cause. It is also a strategy of job search and recruitment that is highly valued by employees and especially by employers as a means of job and employee screening, respectively. We think it unlikely that a federal requirement to list job openings in a national database will substantially transform employers' hiring practices; the kinds of information that both employees and employers want about each other do not appear in standard employment databases, and some employers in our sample distrusted the information provided by job applicants who had no ties to other employees. It seems doubtful that accessibility to a job listing on a computer terminal in a supermarket in one neighborhood will translate into actual accessibility to job openings in other ones.

Because employers are aware of the benefits that follow from a reliance on word-of-mouth recruitment, it seems necessary not only to require a more spatially dispersed advertisement of job listings, but also to alert employers to the ways that word-of-mouth recruitment reproduces and sustains existing sex-based and racial occupational segregation. We see our study as part of this process, and believe that rich spatial stories about geographical containment can serve a transformative pedagogical role by helping employers see the effects of this containment. Recognizing the limits of this type of educational process, however, (insofar as the use of spatially and socially contained networks often benefits employers), we also see the necessity of expanding measures that require employers to open up their recruitment to a wider pool of applicants. To be effective, this will have to involve more than listing jobs in a national (or even a local) registry. Recognizing the significance of personal contacts also underlines the importance of working creatively to expand personal networks among relatively disadvantaged groups and of building bridges to encourage overlapping networks.

Another spatial strategy for disrupting sex-based occupational segregation involves diversifying employment so that a full range of opportunities would be available locally. Our findings suggest that any reordering of the employment landscape so that occupational segregation loses its spatial expression is likely to have the most immediate impact on women with young children working in part-time jobs.

Simply bringing gender-atypical jobs within spatial reach will not by itself, however, ensure the erosion of occupational segregation; the message of importance of place is the need to attend to the range of place-based practices that help to keep occupational boundaries impermeable (e.g., recruitment, scheduling of job hours, spatial organization of work within the enterprise, skill acquisition, counseling about career aspirations). As Wial (1991) and Fernandez Kelly (1994) have noted, workers need more than work "skills" and information about jobs; they need to be able to envision their own place in the labor market.

Finally, our study suggests that evaluations of occupational segregation and of the success and failures of attempts to erase sex-based barriers and boundaries within labor markets must be sensitive to place and to variations in labor market segmentation from place to place. As Wial's (1991) study emphasized, the linkages between workers and jobs do not emerge in some abstract national economy but are forged between specific groups of workers and specific jobs in specific places.[2] Within Worcester, a range of factors come together to make Main South, for example, particularly inhospitable to the crossing of gendered occupational boundaries. We have noted how affirmative action requirements were simply ignored by one employer with whom we spoke. The local culture, the age of the labor force, the local supply of Puerto Rican and Vietnamese workers, the stagnation of local industries all came together to rigidify gendered occupation boundaries. Closer attention to how occupational boundaries get constructed in one place and blurred in another may offer some more general lessons about crossing the boundaries of occupational segregation.

APPENDIX

Given the resources available to us, we initially set out to interview 700 households, chosen so as to be representative of the Worcester-area working-aged population. In the absence of an appropriate sampling frame at the household level, our strategy was to select a probability proportionate to population (PPP) sample in the following manner. First we determined from the 1980 census data the percent of the MSA population living in each town; in addition, to ensure geographic coverage within the city of Worcester, we divided the city area exhaustively into four quadrants and determined the percent of the MSA population in each of these as well. From the 1980 census block statistics we obtained the number of households resident in each census block throughout the MSA. In order to control field costs we decided that, rather than selecting 700 households scattered randomly throughout the MSA, we would aim for clusters of five interviews per census block from 140 randomly selected blocks ($5 \times 140 = 700$).

We then calculated the number of blocks we should sample from each town and from each of the four quadrants within the city of Worcester; this was done by calculating each area's proportion of the MSA population. For example, because 3.5 percent of the Worcester MSA population was living in the town of Holden in 1980, about 3.5 percent of the total 140 sample blocks (or five census blocks) should be drawn from Holden. To select these five blocks at random we took the block statistics and created equal-sized "chunks" of multiples of 40 households. (A block size of 40 households was chosen because this was roughly the modal number of households per block.) Suppose, for example, that adjacent blocks had three, ten, and 27 households on them; these three blocks would be added together to make one "chunk" of 40 households and would be treated in the sampling procedure as one block. The chunks within each town (or area within the city itself) were then selected using a random starting point and a constant interval. If a block had more than 40 households on it, it was weighted accordingly. For example, a block with a cluster of apartment

232

buildings might have 80 or more households on it. Instead of artificially breaking the block into two or more chunks, we simply counted it twice (or more, as appropriate).

Allowing for a 50 percent refusal rate, we initially contacted ten households per block to yield the desired five interviews per block. The households were selected by using a random starting point and a constant sampling interval. If the block had 40 households, we sampled every fourth household, or every eighth if the block had 80 households. The interviewers then approached each targeted household three times on different days of the week and at different time of day in an attempt to make contact with the household. For a number of blocks, however, after three attempts to contact the targeted households, we still could not get five households to agree to be interviewed. Also, the blocks varied in the number of households who were eligible to respond; for example, some blocks had high proportions of elderly households, who were not part of the target population. Therefore, for 39 blocks we have only four households; for eleven we have only three, for four there are only two, and for nine there is only one household per block. We have no sense that these problems introduced any systematic bias to the sample; for example, the blocks with only four households are evenly spread throughout the MSA.

NOTES

1 SPATIAL STORIES AND GENDERED PRACTICES

1 Exceptions are Madden (1981) and Madden and White (1980), who made important early contributions to understanding gender differences in the journey to work.

2 Localities studies emerged from and were mostly confined to the U.K. In the U.K. context, localities are often defined as travel-to-work-areas (TTWAs) which are similar to MSAs in the U.S.

3 The focus of Clark and Whiteman (1983) is on class and race differences rather than on gender.

4 Bowlby *et al.* (1989) locate a reorientation toward feminist rather than geographical sources in the mid-1980s, although their assessment at the end of the decade was that the work of integrating feminist theories and geographies was still "only just beginning" (158).

5 See G. Rose (1993) for a more complete discussion of the links between feminist geographies and contemporary feminist theories.

6 Iris Young argues that there are fundamental differences between masculine and feminine ways of bodily comportment and experiences of space. The two characteristics of feminine comportment and mobility are: the tendency to use a part rather than the whole body when carrying out an action, and the tendency not to extend or stretch into surrounding, available space, in other words, to throw like a girl. Feminine experience of space also differs from the masculine, in three ways. First, Young argues that girls and women tend to be oriented toward enclosed spaces, while boys and men have a spatial orientation that is more open and outwardly directed. Second, girls tend to experience discontinuity between here (where the body is) and beyond. This discontinuity is uncharacteristic of a masculine mode in which the body is unselfconsciously connected to beyond spaces. Third, many girls and women experience themselves as positioned in space in a way that is uncharacteristic for boys and men. In the masculine mode, the body is not lived as an object in space but as an agent that constitutes space. In the feminine mode of spatial existence, one experiences oneself both as object in space and as a subject who constitutes space through action.

2 CONTINUITY AND CHANGE IN WORCESTER, MASSACHUSETTS

1 Worcester's Trade School for Girls was opened in September of 1911; the focus of the Boston researchers' study must, therefore, have been less on

234

whether such a school was justified and more on what the curriculum should be. In 1917 David Fanning, the owner of the Royal Worcester Corset Company, gave the city $100,000 to fund a building for the Girls' Trade School, which was subsequently named for the donor (Nutt 1919: 727–9).

2 The contemporary separation from Boston is exacerbated by Worcester's poor connection to the Massachusetts Turnpike; the only thruway interchange between the Turnpike and Route 290 lies southwest of Worcester in the suburb of Auburn, adding about 15 minutes to the Worcester–Boston trip on the Turnpike. Just why Worcester was not connected directly to the Massachusetts Turnpike when it was built in the 1950s is a matter of some debate. There are tales of a feud between the first chairman of the Massachusetts Turnpike Authority and the editor/publisher of Worcester's *Telegram and Gazette*. Some claim that Worcester's business elite opposed a direct interchange for Worcester on the grounds that it would siphon business away from the downtown. And some claim that difficult terrain and existing neighborhoods made an interchange prohibitively expensive (*Worcester Telegram* 1986; Foster 1986).

3 Later, Worcester native Frances Perkins, would be named Secretary of Labor by President Franklin D. Roosevelt, the first woman to hold a cabinet post.

4 In the 1870 U.S. Census, manufacturing and mechanical industries accounted for 21.6 percent of U.S. employment and fully 54.9 percent of Worcester's. Censuses prior to 1870 do not provide employment data for Worcester City (only for the County).

5 Of course, what is "high-tech" today may be "low-tech" tomorrow. Worcester's wire, envelope, and ceramics industries – considered "traditional" by the mid-twentieth century – had once earned their national and international edge through invention and technological innovation. As testimony to the place-dependent nature of such invention, the process continues, with, for example, the recent development of ceramic wire (Donker 1993).

6 By 1991, about one-third of Massachusetts biotech companies were in the Worcester area (Rosenberg 1991).

7 Mulligan (1980) argues that the wealth and variety of job opportunities that late-nineteenth century industrialization brought to Worcester's women led to their increased independence and willingness to divorce.

8 Data on labor force participation of women with children under 6 are not available for the Worcester MSA for 1960 or 1970; because female labor force participation is now higher in the suburbs than in the city, the MSA figure for 1990 is doubtless closer than is the City figure to the U.S. average.

9 Of course, cities have attracted a disproportionate share of the nation's immigrants. Until recently, the proportions of Worcester's population that is foreign born have been very similar to Boston's: the percentages are, for Worcester and Boston respectively, in 1930, 26 and 29; in 1960, 19 and 15; and in 1990, 8 and 20.

10 In addition to changes in the national and regional sources of immigration to Worcester, the reasons for and experiences of immigration have shifted over time. For experiences of recent U.S. immigrants see Lamphere 1992.

11 A prominent labor leader of the American Federation of Labor, P.J. McGuire, addressing a large crowd of union members in Worcester in 1889, declared that "Worcester has been known for years as one of the scab holes of the state" (Rosenzweig 1983: 20).

12 In Worcester, as elsewhere, unionization rates among women have been

much lower than among men. Only 3.9 percent of Worcester's female workers were union members in 1920, when 17.3 percent of Worcester's male workers belonged to unions. When union membership in Worcester peaked around 1950, 22.3 percent of employed women, compared to 42.4 percent of employed men were in unions. Data on union membership by gender are not available for Worcester for any year after 1951.

13 The local radio station (WTAG) was also owned, for more than 40 years, by the same family until 1987, when the Federal Communications Commission required divestment of dual ownership (of newspaper and radio) in the same market.

14 The concern this change has generated is reflected in a recent headline in the local newspaper, "Ownership Shifting from City." The accompanying article lamented, "Over the last several decades, control of the city's economic base has largely slipped from those who live and work here to people whose lives have been shaped by a different geography. . . . So it has gone, year by year, company by company, store by store, three-decker by three-decker. Ownership of the very things that once defined Worcester as Worcester has passed to people for whom the city's name qualifies as a tongue twister" (Pope 1994: A7).

15 By 1990 the female labor force participation rate was 56 percent in Worcester and 57 percent in the U.S.; the proportion of the labor force that was female (45 percent) was identical for Worcester and the U.S.

3 THE WORCESTER STUDY

1 The difficulties wrought by the lack of data at the appropriate geographical scale are evident, for example, in Stolzenberg and Waite (1984).

2 Cost for the special runs for the Worcester MSA on the 1980 journey-to-work file was about $5,000; Phil Salopek of the Journey-to-Work Division has estimated the cost for doing these runs on the 1990 Worcester MSA would exceed $15,000 (Salopek, personal communication).

3 Occupations were defined as female-dominated, male-dominated, or gender-integrated depending on the proportion of male and female workers in each three-digit census occupation code in 1980. We are aware of the critiques of using occupations instead of job titles to study gender-based labor market segregation (Bielby and Baron 1987, for example, show that jobs within establishment, not occupations, account for most sex segregation; see also Tomaskovic-Devey 1993). These maps would have been impossible to create for job titles, however, as information on job title is not part of the census long form from which the map files were created.

4 Of those 1,132 households who were contacted, found at home, and eligible, 45 percent refused to be interviewed and 55 percent were interviewed. (Note: copies of questionnaires used are available from the authors.)

5 We had both male and female interviewers; 82 percent of the female (and 48 percent of the male) respondents were interviewed by women. We found no significant differences in interview length by sex of respondent or sex of interviewer.

6 The majority (62 percent) of the 151 women who were not in waged employment had chosen not to work outside the home; 21 percent were involuntarily unemployed, 9 percent were retired, and 7 percent were students.

7 Occupations were coded to the three-digit census occupation codes (U.S.

Bureau of the Census 1983). This coding scheme identifies fairly fine occupational divisions. For example, drivers of heavy trucks are distinguished from drivers of light trucks; eleven different types of engineers are specified (e.g., aerospace, mining, civil); short-order cooks are distinguished from other kitchen workers. The census occupation codes are far from perfect, however, especially in the context of identifying gender-based occupational segregation. Waiters and waitresses are, for example, lumped together as are nursing aides, orderlies, and attendants. The level of occupational segregation detected is, of course, extremely sensitive to the level of detail in the occupational coding scheme; in general, the finer the coding, the higher the level of occupational segregation (Bielby and Baron 1984). Further examples of occupations in the census coding scheme, occupations held by people we interviewed, appear in Tables 3.3 and 3.4).

8　A close look at the occupations of men in female-dominated occupations shows that none of these men worked in large female-dominated lines of work like nursing or clerical work and that their being flagged as members of the female-dominated work force is to some extent the artifact of an inadequate occupational coding scheme. The three who were coded as "elementary school teacher," for example, actually taught middle school, for which there is no separate occupation code. One might conclude that in fact the labor force is even more gender-segregated than the presence of only 12 (of 190) men in female-dominated occupations suggests.

9　We did not ask respondents to report their hourly wages. Rather, we asked them to indicate which of the following categories captured their personal annual income: none; less than $5,000; $5,000 to $8,999; $9,000 to $11,999; $12,000 to $14,999; $15,000 to $19,999; $20,000 to $24,999; $25,000 to $29,999; $30,000 to $34,999; $35,000 to $39,999; $40,000 to $49,999; $50,000 to $74,999; $75,000 and over. We calculated the wage variable as $W_i = M_j 50h_i$, where W_i is the hourly wage for person i; M_j is the midpoint of the jth annual income category; h is the number of hours worked per week by person i, and 50 is the number of weeks worked annually.

10　The difference between women in female-dominated occupations and women in gender-integrated ones is small enough to have occurred by chance.

11　All these gender differences in benefits are significant ($p = .000$). The contrast between women in female-dominated occupations and other women is significant for health benefits at $p = .05$, for retirement benefits at $p = .10$, and for both at $p = .10$. Many people have access to these benefits through their spouse or partner. Still, more than two-fifths (43 percent) of the employed women in our sample compared to 35 percent of the men did not have access to a retirement plan through their own job or their partner's. And 17 percent of the women (compared to 13 percent of the men) had no access to health insurance through their job or their partner's.

12　The TSEI assigns an interval-scale score to each three-digit census occupation code in the 1980 census.

13　About 70 percent of both women and men who could identify a job or job title that would be a promotion thought they had a chance of getting that job.

14　More than half (55 percent) of the women with preschool children (under age 6) work part time, compared to only 28.5 percent of those without young children at home ($p = .00$).

15 These figures were calculated from the job histories, which allowed for up to nine entries (ten including current job). Each entry represents a separate job or job title or a time out of the labor force. The average number of years in these work histories is similar for women (16.3) and men (15.7).

16 Women in female-dominated occupations had spent on average 3.5 of the past ten years out of the labor force ($s = 2.9$); for women in gender-integrated occupations the average was 2.8 years ($s = 2.4$); women in male-dominated occupations averaged 3.4 years ($s = 3.0$). The differences are small enough to have occurred by chance ($F = 1.37$, n.s.).

17 We used the special runs from the Journey-to-Work File to reveal the employment makeup (by industry) of each census tract. The location decisions of firms in the health, education, and welfare and the consumer services industries are likely to hinge more on the spatial distribution of potential consumers than on the nature of the local labor force. The location of firms in distributive services is more likely to depend on proximity to transportation facilities and land availability than on labor force characteristics.

18 Using the Coles directory of employers we inventoried all eligible establishments within each of the four studies areas and then telephoned each firm to find out firm type, number of employees, and the name of a contact person. From the master list, we selected firms that were representative of the range of manufacturing and producer services within each area. Firms with fewer than five employees were excluded. We wanted the final sample to reflect the variety of firm sizes and types within each area. The gender composition of the labor force was not a factor in firm selection. Of the 237 firms asked to participate in the study, 159 (67 percent) agreed, and 149 of these were actually interviewed. Ten of the 149 have been excluded from the analysis because they turned out not to be a manufacturing or a producer service firm.

19 The residents of Great Brook Valley whom we interviewed noted that identifying Great Brook Valley as one's place of residence increased the likelihood that employers would not hire them. Similarly, a Great Brook Valley resident told a staff reporter for the *Worcester Telegram and Gazette*, "When you tell a prospective employer you live in Great Brook Valley, you can forget about that job. . . . When you say you live there, you are labelled either a criminal or a drug addict" (Connolly 1989: A2).

20 The Blackstone Valley stretches 46 miles from Worcester to Pawtucket, Rhode Island. Eleven towns comprise the Massachusetts portion of the Blackstone Valley, but we have limited our study to the six Blackstone Valley towns that are within the Worcester SMSA: Grafton, Millbury, Northbridge, Sutton, Upton, and Uxbridge. Several recent histories of the Blackstone Valley include those by Lamphere (1987) and Gerstle (1989).

21 The Valley is sometimes jokingly juxtaposed with the Los Angeles valley of valley girls fame, as in the title of a recent article, "Valley whirl: The Blackstone Valley joins the Massachusetts miracle" (Jones-D'Agostino 1987).

22 Westborough is one of the case studies in Garreau's *Edge City* (1991), a term Garreau uses to describe suburban areas that have experienced intense development of office parks, industrial parks, and shopping malls. Westborough is also included in the "Silicon Valley East" region analyzed

in Saxenian's *Regional Advantage* (1994) in which she compares the industrial culture of California's Silicon Valley with that of the Boston area.

23 Like most New England towns, Westborough retains a town meeting form of government, which places decision making in the hands of those residents who attend town meeting.

24 We thank Michael Brown, Scott Carlin, Glen Elder, Melissa Gilbert, Debby Leslic, Julie Podmore, Suzy Reimer, Stacy Warren, and Michael Zimmer, all research assistants on the Worcester Expedition, for their help with this discussion. We recognize that this process of critical reflection made some former research assistants feel very vulnerable and that some continued to feel exploited by this process; for example, one writes: "Ironically, this exchange [looking back at the Worcester Expedition] continues to promote many of the pitfalls discussed above; in this case, students have now replaced community respondents as the source of information." That the issues of power and knowledge construction persist signals the pressing importance of some of the issues raised.

25 Most of the employers we interviewed expressed an interest in seeing the results of the study, and we did send them a brief report summarizing the findings, as we did for the participants in the 1987 study as well.

26 See Personal Narratives Group (1989) for the different ways that some men and some women have been found to structure their life histories.

27 Rather than framing the academic enterprise through the metaphor of battle, we could reframe it in terms of metaphors of mutual aid or critical vulnerability. (See Lakoff and Johnson, 1980, for a discussion of the ways in which the metaphor of war runs through language used to describe argument – and many other aspects of everyday life.)

4 THE FRICTION OF DISTANCE AND GENDERED GEOGRAPHIES OF EMPLOYMENT

1 Daphne Spain (1992) has also argued that spatial arrangements have been used to keep women from acquiring knowledge.

2 Wajcman (1991) argues that patriarchal urban designers constructed cities to inhibit women's access to the city by, for example, making it difficult to take strollers on public transportation and by including long flights of steps around public buildings (again, discouraging women with children in carriages). In making this point, Wajcman, however, seems to see all women as mothers of young children.

3 Two other studies that we know of (Hanson and Johnston 1985; Singell and Lillydahl 1986) have also documented shorter worktrips among women in female-typed occupations.

4 The question was, "If you were to take another job that paid about the same as the one you have now, what is the most time you would be willing to spend travelling to work."

5 These people are not included in the journey-to-work travel time statistics.

6 These maps were created from people in the 1987 sample. Because only four men were interviewed in Upper Burncoat, the median distance for men living in this area is not included.

7 The Pearsons r for the relationship between percent of a tract's employment that is in female-dominated occupations and the percent that is in male-dominated occupations is −0.82.

8 Maps of the other industry categories listed in Table 3.2 show the same general pattern of spatial gender divisions within industry.

9 Madden and Chiu (1990) examine the spatial variation in wages over 23 zones in the Detroit metropolitan area and 27 zones in the Philadelphia area; they show that the interzone variability in wages in both cities is higher for women than for men, implying that the wage consequences of spatial constraints are greater for women (360).

10 Others have sought to expand measures of local context to include indices of child care availability, but this too has been for very large geographical units, such as county groups (Stolzenberg and Waite 1984). Ashton *et al.* (1988) document the profound impact of local labor markets on labor market outcomes for youths in the United Kingdom.

11 The median distances were calculated from the sample: for women working full time and living in the city, 2.86 miles; working full time and living in a suburb, 3.73 miles; working part time and living in the city, 1.86 miles; working part time and living in a suburb, 2.86 miles.

12 For example one group is comprised of women with preschool children, greater-than-high-school education, and full-time jobs. Another includes women without preschool children, greater-than-high-school education, and part-time jobs.

13 Similar in the sense that they all have more than a high school education, have small children, and work part time.

14 In the discriminant analysis of part-time employed women with children under age 6, there were 33 women in female-dominated occupations and 15 in other occupations. The discriminant function was able to classify 79 percent of the cases successfully; it correctly identified 81 percent of the women in female-dominated occupations and 73 percent of the women in other occupations. The Wilks' Lambda for the function is statistically significant ($p = .04$). In the analysis of part-time employed women with more than a high school education and children under age 6, there were 19 women in female-dominated occupations and nine in others. The discriminant function was able to classify 82 percent of the cases successfully; it correctly identified 74 percent of the women in female-dominated occupations and all of the women in other occupations. The Wilks' Lambda is significant ($p < .01$).

15 Part-time employed mothers of pre-school children had an average commute time of 15.2 minutes (median 10 minutes); although the average for full-time employed mothers of young children is not different from their part-time counterparts (15.4 minutes), the median for this latter group is 15 minutes.

5 HOUSEHOLD ARRANGEMENTS AND THE GEOGRAPHY OF EMPLOYMENT

1 The term "strategies" has a contested history in recent studies of the household (Laurie and Sullivan 1991; Jordan *et al.* 1992). When we use the term we do not mean to imply long-term rational planning. Nor do we assume that strategies are always conceived cooperatively or result from equal power relations within households. The term does imply that households work as a type of unit, and that the life and work of one member is intrinsically connected to that of others in the household. It also suggests

that members of households play active roles in their lives, choosing one way of living (consciously or not) among others.

2 Our focus here is on patriarchal relations within the family. These are created within, and buttressed by, patriarchal policies of the state, such as those surrounding welfare payments and child care (see, for example, Fraser 1989; Fincher 1993).

3 By 1990, according to the U.S. Census, the gap between Worcester and the U.S. had narrowed, with 41 percent of the Worcester MSA population and 46.7 percent of the U.S. population living in another residence five years previously.

4 We have so few nonwhite respondents in our sample that it is difficult to draw systematic conclusions in terms of race. However, given that so many Latino and Vietnamese residents have recently arrived in Worcester, our generalizations about residential stability can hardly apply to them.

5 Here we are using the manual employment of male household heads to identify working-class households. We recognize the conceptual and strategic difficulties of allocating household class position on the basis of the occupation of the male household head (Pratt and Hanson 1991). When forced to allocate class positions to households (rather than individuals) we nevertheless have used male occupation. Given the existing wage differential between men and women in dual-headed households, it is typically the male occupation that affects differential resources across households, such as those enabling a household to live in a working- or middle-class neighborhood (Pratt and Hanson 1988). In the specific case of the intergenerational transfer of wealth, we highlight male occupations because property was more often transferred through the male line (see Pratt and Hanson 1991). This housing strategy was less common among households of other classes: again classifying households in terms of the occupational standing of male heads of household, only 5 percent of nonskilled nonmanual, 10 percent of skilled nonmanual and 6 percent of managerial and professional households enjoy this type of transfer of wealth.

6 Though we did not ask about this explicitly in the interviews, five female respondents volunteered the information that inheriting a house had allowed them to work only part-time in the labor force. Employed women in households living in housing obtained from a family member were significantly more likely to be working outside the home part time as opposed to full time; 41 percent of the employed women in such households versus 33 percent of employed women in other households were working part time ($p = .10$).

7 Among never-married women in this occupational category, fully 71 percent find their jobs first and then find their residence. For married women in other occupational groupings, only 6.5 percent of those in skilled nonmanual, 2.1 percent in nonskilled nonmanual, and none in manual occupations found their residence after their job.

8 We asked people to say how important each of a number of factors was in their decision to live in their current house or apartment and then to identify the three factors that were most important in that decision. Choosing a residence that was close to work was among the three most important for more than a third (35 percent) of the men we interviewed but less than one-sixth (14 percent) of the women. Men, then, are much more likely than women to see locating their residence close to their workplace as

241

a possibility and to place a high priority on this consideration in residential choice.

9 Women did not make this same distinction: about half thought that convenience to their own and their partner's jobs was important or very important to their choice of current residence (51 and 55 percent, respectively).

10 53 percent felt that the family would move if their husbands could get a better job, but only 23 percent thought that an improvement in their own job would prompt a household move.

11 Only 13.5 percent of women in female-dominated occupations said they would move for a better job, while 27.8 percent of women in male-dominated occupations and 23.5 percent of women in gender-integrated occupations were willing to undertake a move for a better job.

12 Gender-typing of occupations is more clearly related to women's willingness to move than is their occupational class: one quarter (23 percent) of married women in managerial or professional occupations would consider moving, but so too would 38 percent of women in nonskilled manual occupations. It seems likely that there is a very different dynamic involved in each case.

13 Women with a college education are significantly less likely ($p = .001$) to have taken a child-related break than are women with less education, a pattern also found by Jacobsen and Levin (1992) and Desai and Waite (1991). Only 57.9 percent of college-educated mothers in our sample had taken a child-related break, compared to 75.8 percent of mothers with only a high school diploma. To compare the entry-level employment characteristics of women who did and did not take breaks, we calculated the occupational prestige score (the Total Socio-Economic Index – TSEI) for each woman's first job. The average first-job TSEI for mothers who had taken a family-related break was significantly lower ($p = .02$) than the average first-job TSEI score of mothers who had not taken a family-related break.

14 We measured this both by TSEI score ($F = 1.25$, $p = .26$) and occupational class (e.g., nonskilled nonmanual, etc.) ($p = .66$).

15 Fully 75 percent of women in middle-income households had taken a childcare break as opposed to roughly 50 percent in other households.

16 As we noted in Chapter 3, among women in our Worcester study, having a job in a female-dominated occupation in 1987 was not related to the number or lengths of breaks a woman had taken from paid employment.

17 Likelihood of shortening commute by at least 10 minutes was not associated with any of the following variables: current occupational class, occupational class preceding break, occupational class of job immediately following break, current personal income, family income, occupational type (e.g., female-dominated), family circumstances (e.g., divorced, married), woman's educational attainment, number of children in the household.

18 We compared the TSEI of job held immediately before and immediately after the childbearing break (for women with more than one break we used the most recent one). We defined those experiencing upward or downward mobility as those with an increase or decrease of ten or more in the TSEI, respectively. As another way of stating the results of the analysis, we found that 58 percent of women whose return to the labor force entailed a downward shift in occupational status returned to jobs that were also at least ten minutes closer to home; by contrast, only 36 percent of the women experiencing upward mobility (and 25 percent of those with no appreciable

change in occupational status in jobs they held before and after a break) reduced their commutes by as much as ten minutes ($p = .06$).

19 Among women married to managers or professionals, 44 percent worked part time when interviewed, compared to 42 percent married to skilled nonmanual workers, 40 percent married to skilled manual, 20 percent married to nonskilled nonmanual, and 29 percent married to nonskilled manual workers. It is also the case that women married to professionals and managers were the group who most wanted part-time work. Of those women who worked full time, almost half (43.2 percent) of those married to men in managerial and professional occupations would prefer to work part time, compared to 31.7 percent married to men in skilled and non-skilled nonmanual occupations and 23.9 percent of women married to men in nonskilled and skilled manual occupations. These latter differences are not, however, statistically significant ($p = .31$).

20 Of women working in part-time female-dominated jobs, 87.7 percent thought that being close to home was an important or very important job attribute. But this was equally true for those in other occupations: 78.4 percent of those in part-time gender-integrated and all of those in part-time male-dominated jobs felt this way.

21 For reasons that are not obvious to us, most of the women who desired this change were in gender-integrated jobs.

22 Given that gender-typing of occupations intersects with occupational class, we also find that there are many more part-time workers among "lower" occupational grades. Almost half of the women working in nonskilled manual and nonmanual occupations (41.9 percent and 46.1 percent respectively) work part time. Roughly one-third (30.0 percent) of women working in skilled manual and nonmanual occupations work part time and only 26.7 percent of women working in managerial and professional occupations work part time.

23 We measured mobility by calculating the average change in TSEI (occupational status) score divided by years in the labor force. Average annual mobility for women presently working full time was 0.31 while the average for part-time workers was -0.004 ($p < .12$).

24 The use of this strategy extends across racial categories: of those arranging their schedules sequentially, 88 percent were identified as "white," 3 percent as "black," 1.4 percent as "Hispanic," and 7 percent as "other," proportions that roughly parallel those in the sample as a whole. We should note that the usual hours during which stores are open in the Worcester area (9 a.m.–9 or even 10 p.m., including Saturdays and Sundays) enables those who work at "odd" hours to combine paid work with household work.

25 In the case of child care, the man received a point if he only assisted with child care.

26 It is worth noting that the gender of occupation is a better "predictor" of household division of labor than are the characteristics that are frequently cited in the existing academic literature. Just as Ericksen et al. (1979: 310) report (cited in Morris 1990: 99), men with higher personal incomes are less likely to help with domestic work ($p = .01$), but we find no relationship between a woman's own or family's income and her husband's likelihood of helping with domestic work. We also find no support for the argument that role segregation between husband and wife increases with the con-nectedness of the family's social networks. There was, for example, no relationship between numbers of friends and kin in one's neighborhood

and role segregation with the household. (For a review of other recent studies that pursue this hypothesis, see Morris 1990.)

27 This relationship is even stronger for a subsample of households with small children (5 years or younger). In these households, 50 percent of women in male-dominated occupations reported that their husbands took responsibility for at least two household tasks, as compared to 12.0 percent in gender-integrated and 9.0 percent in female-dominated occupations ($p = .002$).

28 Men with working wives who report no involvement in domestic work have longer average travel times to work (24.4 minutes, $s = 18.6$) than those with involvement (20.9 minutes; $s = 15.1$, $p = .09$). For men whose wives are not employed, participating in domestic work has no impact on their travel time to work. For men whose wives are employed, however, contributing to household chores is related to having a shorter journey to work.

29 The number of mature women in male-dominated jobs is so low (only nine) that the results are not statistically significant. However, only 33 percent of women in male-dominated occupations had Household Responsibility Index (HRI) scores of three or four, as compared to 55.1 percent of women in female-dominated occupations and 53.2 percent in gender-integrated ones. On the other hand, fully 44.4 percent of women in male-dominated occupations had HRI scores of 0 or 1, and only 13.1 percent of women in female-dominated and 19.1 percent in gender-integrated occupations had scores in this range.

30 The never-married group of women differ from others in many ways: they are younger (average age of 34.8), better educated and typically without children. The divorced, widowed, and separated women were quite similar to the married women with respect to a number of social characteristics, making the discrepancy in employment experience all the more interesting. Their ages were fairly similar: an average of 44.1 for the divorced, widowed, and separated women and 40.1 percent for the married ones. They have similar numbers of children living at home. Among married women, 26 percent had no children, and 53 percent had one to two children living at home. Considering the divorced, separated, and widowed women, 23 percent reported no children at home, and another 57.1 percent had from one to two children living with them. Their educations are roughly comparable, although there is a slight (but not statistically significant) tendency for married women to be better educated. This is an interesting direction of difference, given that married women are more likely to find employment in female-dominated occupations. Considering married women: 8 percent had less than high school, 38.4 percent had a high school diploma, 5.8 percent had attended a technical or secretarial program, 38.9 percent had some college or a college degree, and 7.5 percent had some postgraduate college education. Among those divorced, separated, or widowed; 17.2 percent had no high school diploma, 36.2 percent had only a high school diploma, a further 7.2 percent attended a technical or secretarial program, 33.2 percent had attended college, and 5.7 percent had some postgraduate college education.

31 Not surprisingly, this does not translate into higher family incomes among households headed by women. Only 3.3 percent of married women lived in households with incomes lower than $9,000 while fully 24.5 percent of divorced, separated, and widowed households and 11.1 percent of single households did. Alternatively, 31.1 percent of married women lived in

households with incomes of $35,000 or more while only 15.1 percent of divorced, separated, and widowed and 11.1 percent of single women had this family income.

32 We asked interviewees to indicate whether they felt that a particular job attribute was very important, important, neither important nor unimportant, unimportant or very unimportant. Of divorced, separated, and widowed women, 57 percent felt that "possibilities for advancement" were important or very important. The comparable percentages for never-married or married women are 62 percent and 34 percent, respectively ($\chi^2 = 16.3$, $d.f. = 4$, $p = .002$). For the attribute "good benefits," a similar proportion of divorced, separated, and widowed, and never-married women (53 percent and 54 percent, respectively) felt that this was important or very important but only 39 percent of married women felt this way ($\chi^2 = 12.72$, $d.f. = 4$, $p = .01$).

33 To put this into perspective, an income of $35,000 or more was not rare among men in Worcester; more than 34 percent of men with whom we spoke were in this income bracket.

34 It is worth underlining our finding that divorced women tend to find employment very close to home in light of England's (1993) discovery that this was not the case among a nonrandom sample of divorced clerical workers in Columbus, Ohio. It is important to note this discrepancy in empirical findings because England takes her finding as evidence against the "conventional wisdom" that divorced women are among the most spatially constrained.

35 This compares to 24 percent of married women, 14 percent of married men, and 20 percent of women and no men who were divorced, separated, and widowed.

36 The numbers are so small that the relationship between caring for others and occupational type for these groups of women is not statistically significant. Nevertheless, combining divorced, separated, widowed and single women, 26.9 percent of women in female-dominated and gender-integrated occupations reported caring for someone other than a child or partner on a regular basis, compared to 0 percent in male-dominated occupations.

37 The *National Study of Changing Workforce* (Galinsky *et al.* 1993) also documented – with some surprise – the significance of men's care-giving responsibilities for elders. Although only 7 percent of the workers surveyed in that study cared for disabled spouses, relatives, or friends who were 50 years or older, almost half of these caregivers (44 percent) were men. Galinsky *et al.* note: "The common assumption that women are far more likely to have elder care responsibilities is erroneous" (p. 58). They do find, however, that women devote significantly more *time* to elder care than do men.

6 EMPLOYER PRACTICES, LOCAL LABOR MARKETS, AND OCCUPATIONAL SEGREGATION

1 We are, of course, relying on employers' recollections of their reasons for locating; the indeterminate relationship between reasons for acting and *ex post* rationalization is one that plagues much survey research.

2 Others (e.g., Vipond 1984; Rees and Shultz 1970) have pointed out that

employers prefer workers who live close by and screen applicants on distance from the workplace.

3 In Main South, in particular, a large proportion of people do not have access to a car. The 1980 census indicates that fully 41 percent of households resident in Main South have no vehicle, and 29 percent of employees interviewed in Main South in 1989 did not have access to a car. Only 9 percent of women employees in the Blackstone Valley and 10 percent in Upper Burncoat reported having no access to a car.

4 Some employers recognized the problems that can attend this strategy of labor recruitment. One employer who hired through family networks said that he tried not to hire too many members of the same family because if there is a family emergency "the whole line shuts down." Others recognized the point that Manwaring (1984) and Granovetter and Tilly (1988) have highlighted: strong social networks, fed by word-of-mouth recruitment and long job tenure, build worker solidarity and strengthen workers' bargaining positions. One employer expressed this recognition in the following way: "We have tried to avoid having relatives on the same shift. Family feuds come into this business. If you discipline one of them, the whole family is mad at you."

5 For an analysis of Norton's recent labor history see Jonas (1992).

6 At the time of the interview, the move had not been completed.

7 Half of the firms that had moved had located within 10 km (6 miles) of their previous location, and two-thirds had moved no more than 23 km (14 miles).

8 Twenty-five firms indicated that they planned to move within two or three years. All but two were planning to move within the Worcester area and half were planning to relocate within the same small area of their current location.

9 We do not know the comparable figure for the owners we did not interview.

10 Other studies of the segmented labor process within workplaces have also documented the barriers to communication across job types (e.g., Stull *et al.* 1992; Grenier *et al.* 1992).

7 COMMUNITIES, WORK, AND GENDER RELATIONS

1 Only one-third of these households had moved more than 5.5 miles from their previous residential location to their current homes.

2 Of those who were from Worcester, 53 percent had found their current residence informally, but only 43 percent of those who were not, had done so ($p = .01$). And nearly four-fifths (79 percent) of those who had grown up in Worcester, compared to "only" 72 percent of those who had not, had found their present jobs through informal channels ($p < .04$).

3 These personal networks also exert a centripetal tug on Worcester natives who have moved away. A 40-year-old single woman, for example, who had grown up in Worcester and whose family was still there, had reluctantly moved away to pursue her career as a junior-high school teacher and had eagerly watched for an opening in Worcester so she could return. Originally unable to find a teaching position in Worcester but not wanting to move away, she had done clerical work for the first five years after earning her teaching credentials and then had moved to a teaching post in Connecticut for five years. Currently living with her parents in a Worcester suburb, she told us, "I waited ten years to get a job in Worcester. My family is here and I

wanted to work here. I got experience in Connecticut and waited until a position became available here."

4 We distinguished between contacts or information sources that people said they had used in finding housing and those through whom they had actually found their current house or apartment.

5 Although people who have not grown up in Worcester made significantly less use of relatives in finding housing, fully 30.4 percent of this group had received information from family members and 15 percent had found their current house or apartment through a relative.

6 Less than 9 percent of the people in our sample had commutes longer than 35 minutes.

7 49 percent of nonskilled nonmanual and 42.5 percent of skilled nonmanual had used informal channels (p = .000).

8 These figures are in line with those reported by Montgomery (1991) in his review of the literature on job finding, in which he finds that about half of all workers obtain their jobs through personal contacts. We asked respondents whether or not they had actively searched for their present jobs, and then, regardless of whether they had actively searched or not, we asked them to describe, in an open-ended way, the circumstances under which they had found their current employment. The issue of what constitutes an active search was left open; our question was aimed to tap the respondents' own perceptions of whether or not they had engaged in active search in obtaining their present jobs. Respondents were reminded that active search could take place when one is employed.

9 This was the case for 75 percent of women and 79 percent of men. Considering only personal contacts, 54 percent of women and 50 percent of men found their present job through a personal contact. For women, job information obtained through personal contacts was usually a substitute for a formal job search: 78 percent of women who found their jobs without a search used personal contacts while only 25 percent of active searchers had used a personal contact to get their jobs. For men, the use of personal contacts is more often part of a formal job search: 59 percent of those who did not search and 42 percent of those who actively searched obtained their jobs through personal contacts.

10 Finding a job informally includes finding a job through personal contact, seeing an ad in the window, direct application, and being recruited.

11 Most of those known before were relatives: 27 percent of all employees interviewed had a relative at work now, and 15 percent had a friend working there now whom they had known before.

12 11 percent of those with no relative and 13 percent of those with no friend at work now had had one there in the past.

13 Our discussions with this woman were conducted in Spanish. What follows is a translation of what she told us in Spanish.

14 Moore (1990) has argued that these gender differences in the composition of personal networks are due to differences in women's and men's structural positions with respect to employment, marriage, and children; when women's and men's positions in society become similar, the composition of their personal networks will converge.

15 The family contacts linking both men and women to jobs are more likely to be male than female: 58 percent of women's family contacts and 67 percent of men's family contacts were a male relative.

16 As an interesting historical footnote, Esther Howland, the woman who in 1848 began the production of valentines in Worcester and who over thirty years built a successful valentine enterprise, did so with the encouragement and financial backing of her father who was a stationer (Schantz 1991). See

Schroedel (1985) for additional anecdotal evidence on the importance of male family members in women's entrance into nontraditional occupations.

17 Jordan *et al.* (1992) also highlight the importance of community contacts for job information among women and men earning low incomes. See also Whipp (1985), Grieco (1987), and Wial (1991).

18 We suspect that these results would be strengthened considerably if we had information about the large numbers of "unknown" contacts.

19 Almost half (45 percent) of Blackstone Valley residents interviewed in 1987 grew up in the Worcester area; the average length of residence in Worcester was 25.3 years; and half of all women interviewed reported having at least one relative living in their neighborhood.

20 In Westborough 30.3 percent of women had at least one university degree, compared to 12 percent in Main South, 16.7 percent in Upper Burncoat, and 14.3 percent in the Blackstone Valley. Only 6.2 percent of women in Westborough had no high school diploma, while this was true for 28 percent interviewed in Main South, 27.7 percent in Upper Burncoat, and 15.7 percent in the Blackstone Valley (these statistics are compiled from 1987 interviews).

21 For an assessment of women's success in breaking into the occupations of computer systems analyst, see Donato's case study in Reskin and Roos (1990). Donato's assessment is that women's progress in this occupational category has "not been unequivocal" (187): women are still segregated into lower-paid specialties in the occupation.

22 For this analysis, we examined the occupations that were classified as male-dominated according to the U.S. 1980 census. Occupations were classified as male-dominated if at least 70 percent of incumbents were men. According to this criterion, even in the areas where we note a greater proportion of women in male-dominated occupations (e.g., 21.5 percent of managers in Westborough are women) the numbers of women would not lead to a reclassification of the occupation as gender-integrated. In only two cases – sales of manufacturing goods in Upper Burncoat and machine operators in metal, plastics, and lathe in Westborough – could a traditionally male occupation be said to be gender-integrated.

23 In comparison, 27 percent of employers in Westborough who run evening or night shifts said that they do not hire women because women do not want this schedule. 13 percent of employers in Upper Burncoat and 40 percent of Blackstone Valley employers who offer night shift work also gave this reason for not hiring women to work the night or evening shift.

24 Few women living in the other areas reported that they had no access to an automobile. This was the case for only 6 percent in Westborough, 10 percent in Upper Burncoat, and 9 percent in the Blackstone Valley.

25 We made pairwise contrasts between the average wages in each area. Four areas and seven job types made for 42 tests of significance. All but 12 of these proved statistically significant at $p = .01$. For *general unskilled*, all pairwise contrasts of average wages between areas are significant at $p = .01$ except that between Upper Burncoat and the Blackstone Valley. For *routine production*, all pairwise contrasts are significant at $p = .01$ except that between Main South and the Blackstone Valley. For *skilled production/routine supervisors* all pairwise contrasts are significant at $p = .01$. For *engineers*, all pairwise contrasts are significant at $p = .01$ except those between the Blackstone Valley and Main South; and the Blackstone Valley and Upper Burncoat. For *clerical/office* jobs, all pairwise contrasts are significant at

$p = .01$ except those between Main South and the Blackstone Valley, and Main South versus Upper Burncoat. All *sales/marketing* contrasts are non-significant, except those between Main South and Westborough, which are significant at $p = .01$. For *professional/managers*, all contrasts are significant at $p = .01$ except those between Upper Burncoat and the Blackstone Valley.

26 Ihlandfelt and Young (1994) have documented similar spatial wage differences in Atlanta. They examined wages in fast-food restaurants as a function of distance from downtown Atlanta and found a positive wage gradient: workers in suburban establishments earned significantly more for the same work.

27 See Reskin and Roos (1990) for a discussion of the ghettoization of secretarial/editorial assistant occupations in the secondary labor market.

28 We used the median for all city tracts and the median for all suburban tracts as the dividing line between high- and low-income tracts.

29 For women living in the City of Worcester, there was no statistically significant difference between pre- and post-break commute: 15.07 minutes and 13.36 minutes, respectively.

30 There is no relationship between shortening commute and occupational mobility for women living in lower-income suburbs.

8 CROSSING BOUNDARIES

1 Myers (1970: 191) fully recognized the limits of a purely accessibility-based approach to alleviating inner-city unemployment and poverty: "While transportation is often a necessary condition for the accomplishment of worthy social objectives, rarely is it both necessary and sufficient." Moreover, Myers recommended that a private, rather than government, agency should run the "new Volks for poor folks" program: "given how conservatives feel about the poor and how radicals feel about the automobile, I have trouble imagining a government program such as I have described" (202).

2 This point suggests to us that, when an old industry dies because it can no longer compete, and the local labor force remains largely in place (as it usually does), it is *local* entrepreneurs, with their extensive knowledge of local skills and practices, who should be in the best position to launch new ventures.

REFERENCES

Abrahamson, M. and Sigelman, L. (1987) "Occupational sex segregation in metropolitan areas", *American Sociological Review* 52: 588–97.

Allen, K. (1984) *On the Beaten Path: Westborough, Massachusetts*, Westborough: Westborough Civic Club and Westborough Historical Society.

Anonymous (1975) "Employment data: unemployment rates by sex and age, monthly data seasonally adjusted", *Monthly Labor Review* 98: 32.

Ashton, D., Maguire, M., and Spilsbury, M. (1988) "Local labour markets and their impact on the life chances of youths", in B. Coles (ed.) *Young Careers: The Search for Jobs and the New Vocationalism*, Milton Keynes: Open University.

Bagguley, P., Mark-Lawson, J., Shapiro, D., Urry, J., Walby, S., and Warde, A. (1990) *Restructuring: Place, Class and Gender*, London: Sage Publications.

Barff, R. (1990) "The migration response of the economic turnaround in New England", *Environment and Planning A* 22: 1497–516.

Barrett, M. (1980) *Women's Oppression Today*, London: Verso.

Barrett, M. and McIntosh, M. (1982) *The Antisocial Family*, London: Verso.

Barron, J., Bishop, J., and Dunkeberg, W. (1985) "Employer search: the interviewing and hiring of new employees", *The Review of Economics and Statistics* 67: 43–52.

Bartky, S.L. (1988) "Foucault, femininity, and the modernization of patriarchal power", in I. Diamond and L. Quinby (eds.) *Feminism and Foucault*, Boston: Northeastern University Press.

Bateson, M.C. (1990) *Composing a Life*, New York: Plume.

Beechey, V. (1977) "Some notes on female wage labor in capitalist production", *Capital and Class* 3: 45–66.

Beechey, V. and Perkins, T. (1987) *A Matter of Hours*, Cambridge: Polity Press.

Beller, A.H. (1984) "Trends in occupational segregation by sex and race, 1960–1981" in B. Reskin (ed.) *Sex Segregation in the Workplace: Trends, Explanations, Remedies*, Washington D.C.: National Academy Press.

Bielby, W.T. and Baron, J.N. (1984) "A woman's place is with other women: sex segregation within organizations", in B. Reskin (ed.) *Sex Segregation in the Workplace: Trends, Explanations, Remedies*, Washington D.C.: National Academy Press.

—— (1987) "Undoing discrimination: job integration and comparable worth", in C. Bose and G. Spitze (eds.) *Ingredients for Women's Employment Policy*, Albany: SUNY Press.

Blackburn, R.M. and Mann, M. (1979) *The Working Class in the Labour Market*, London: Macmillan.

250

REFERENCES

Blewett, M.H. (1989) *Men, Women, and Work: Class, Gender, and Protest in the New England Shoe Industry, 1780–1910*, Illinois: University of Illinois Press.

Bluestone, B. and Harrison, B. (1982) *The Deindustrialization of America*, New York: Basic Books.

Bondi, L. (1992) "Gender and dichotomy", *Progress in Human Geography* 16: 98–104.

Bowlby, S., Lewis, J., McDowell, L., and Foord, J. (1989) "The geography of gender", in R. Peet and N. Thrift (eds.) *New Models in Geography*, Vol. 2, London: Unwin and Hyman.

Brannen, J. (1989) "Childbirth and occupational mobility: evidence from a longitudinal study", *Work, Employment, and Society*, 3: 179–201.

Browne, L. (1988) "Defense spending and high technology development: national and state issues", *New England Economic Review* September/October: 3–22.

Budz, J. (1978) "Employers' view of the labor market and job opportunities for working women in the Worcester area", Mimeo, Worcester Area Career Education Consortium.

Case, K. (1992) "The real estate cycle and the economy: consequences of the Massachusetts boom of 1984–87", *Urban Studies* 29: 171–83.

Cho, S.K. (1985) "The labor process and capital mobility: the limits of the new international division of labor", *Politics and Society* 14: 185–222.

Christopherson, S. (1983) "The household and class formation: determinants of residential location in Ciudad Juarez", *Environment and Planning D: Society and Space* 1: 323–38.

——— (1989) "Flexibility in the US service economy and the emerging spatial divison of labor", *Transactions of the Institute of British Geographers* 14: 131–43.

City of Worcester (1991) *Profile of Worcester, Massachusetts*, Massachusetts: City of Worcester.

Clark, G.L. (1983) "Fluctuations and rigidities in local labor markets, parts I and II", *Environment and Planning A* 15: 165–85, 365–77.

Clark, G.L. and Whiteman, J. (1983) "Why poor people do not move: job search behaviour and disequilibrium amongst local labour markets", *Environment and Planning A* 15: 85–104.

Cobble, D.S. (1991) " 'Drawing the line': the construction of a gendered work force in the food service industry", in A. Baron (ed.) *Work Engendered: Toward a New History of American Labor*, Ithaca: Cornell University Press.

Cockburn, C. (1983) *Brothers: Male Dominance and Technological Change*, London: Pluto.

Cohen, B. (1988) "The Worcester machinists' strike of 1915", *Historical Journal of Massachusetts* 10: 154–71.

Commonwealth of Massachusetts (1921) *Annual Report on the Statistics of Labor*, Boston: Department of Labor and Industries.

Commonwealth of Massachusetts, Department of Public Health (1992) *Rates of natural population increase in Massachusetts, 1900–1989*, Massachusetts: The Commonwealth of Massachusetts, Department of Public Health.

Connolly, T. (1989) "GBV 'labels' are noted at hearing", *Worcester Telegram and Gazette*, 31 May: A2.

Cooke, P. (1983) "Labour market discontinuity and spatial development", *Progress in Human Geography* 7: 543–66.

Corcoran, M., Datcher, L., and Duncan, G. (1980) "Information and influence networks in labor markets", in G. Duncan and J.N. Morgan (eds.) *Five*

Thousand American Families: Patterns of Economic Progress Volume 8, Ann Arbor: Institute for Social Research, University of Michigan.

Corcoran, M., Duncan, G.J., and Ponza, M. (1983) "Work experience and wages: growth of women workers", in G. Duncan and J.N. Morgan (eds.) *Five Thousand American Families: Patterns of Economic Progress* Volume 10, Ann Arbor: Institute for Social Research, University of Michigan Press.

—— (1984) "Work experience, job segregation, and wages", in B. Reskin (ed.) *Sex Segregation in the Workplace: Trends, Explanations, Remedies*, Washington D.C.: National Academy Press.

Cosgrove, D. (1985) "Prospect, perspective and the evolution of the landscape idea", *Transactions of the Institute of British Geographers* 10: 45–62.

Cox, K. and Mair, A. (1988) "Locality and community in the politics of local economic development", *Annals of the Association of American Geographers* 78: 307–25.

Crick, M. (1992) "Ali and me: an essay in street-corner anthropology", in J. Okely and H. Callaway (eds.) *Anthropology and Autobiography*, London: Routledge.

Cunnison, S. (1987) "Women's three working lives and trade union participation", in P. Allatt, T. Keil, A. Bryman, and B. Bytheway (eds.) *Women and Life Cycle: Transitions and Turning Points*, London: Macmillan.

Davis, K. (1988) "Wives and work: a theory of the sex-role revolution and its consequences", in S.M. Dornbusch and M.H. Stroker (eds.) *Feminism, Children, and the New Families*, New York: Guilford Press.

de Certeau, M. (1984) *The Practice of Everyday Life*, Berkeley: University of California Press.

de Lauretis, T. (1988) "Displacing hegemonic discourses: reflections on feminist theory in the 1980s", *Inscriptions* 3/4: 127–44.

—— (1989) "The essence of the triangle or, taking the risk of essentialism seriously: feminist theory in Italy, the U.S. and Britain", *Differences* 1: 3–37.

Desai, S. and Waite, L.J. (1991) "Women's employment during pregnancy and after the first birth: occupational characteristics and work commitment", *American Sociological Review* 56: 551–66.

Deutsche, R. (1991) "Boys town", *Environment and Planning D: Society and Space* 9: 5–30.

Dex, S. (1987) *Women's Occupational Mobility: A Lifetime Perspective*, New York: St. Martin's Press.

Diesenhouse, S. (1991) "Downturn in Boston area is expanding to Route 495", *New York Times* 23 January: D18.

Doane, M.A. (1987) *The Desire to Desire*, Bloomington: Indiana University Press.

Donker, P. (1993) "Company wired to future", *Worcester Telegram and Gazette* 20 August: D1.

Duffy, A., Mandell, N., and Pupo, N. (1989) *Few Choices: Women, Work and Family*, Toronto: Garamond Press.

Eastman, R. (1992) *IDRISI*, Worcester, Massachusetts: Clark University Press.

Emberley, P. (1989) "Places and stories: the challenge of technology", *Social Research* 5: 741–85.

England, K. (1993) "Suburban pink collar ghettos: the spatial entrapment of women?", *Annals of the Association of American Geographers* 83: 225–42.

England, P. (1982) "The failure of human capital theory to explain occupational sex segregation", *Journal of Human Resources* 17: 358–76.

England, P. and Farkas, G. (1986) *Households, Employment, and Gender: A Social, Economic, and Demographic View*, New York: Aldine.

252

REFERENCES

Ericksen, J.A. (1977) "An analysis of the journey to work for women", *Social Problems* 24: 428–35.

Erskine, M. (1981) *Worcester: Heart of the Commonwealth,* Woodland Hills, CA: Windsor Publications.

Fagnani, J. (1983) "Women's commuting patterns in the Paris region", *Tijdschrift voor Economische en Sociale Geographie* 74: 12–24.

Feingold, N. (1986) *Women's Work: the Worcester Experience,* Worcester, MA: The Worcester Historical Museum.

Ferguson, K. (1983) *The Man Question: Visions of Subjectivity in Feminist Theory,* Berkeley: University of California Press.

Fernandez Kelly, M.P. (1994) "Towanda's triumph: social and cultural capital in the transition to adulthood in the urban ghetto", *International Journal of Urban and Regional Research* 18: 88–111.

Fincher, R. (1993) "Women, the state, and the life course in urban Australia", in C. Katz and J. Monk (eds.) *Full Circles: Geographies of Women over the Life Course,* New York: Routledge.

Fine, L. (1990) *The Souls of the Skyscraper: Female Clerical Workers in Chicago, 1870–1930,* Philadelphia: Temple University Press.

Fishman, R. (1990) "Megalopolis unbound", *The Wilson Quarterly,* Winter: 25–48.

Flax, J. (1990) *Thinking Fragments: Psychology, Feminism, and Postmodernism in the Contemporary West,* Berkeley and Los Angeles: University of California Press.

Forer, P.C. and Kivell, H. (1981) "Space–time budgets, public transport, and spatial choice", *Environment and Planning A* 13: 497–509.

Foster, D. (1986) "You can't get there from here: The 146 connector", *Worcester Telegram* 18 March: 15–17.

Foucault, M. (1990) *The History of Sexuality, vol. 1,* New York: Vintage Books. Translated by Robert Hurley.

Fraser, N. (1989) *Unruly Practices: Power, Discourse, and Gender in Contemporary Social Theory,* Minneapolis: University of Minnesota.

Freeman, R.B. and Medoff, J.L. (1979) "New estimates of private sector union-ism in the United States", *Industrial and Labor Relations Review* 32: 143–74.

Fuss, D. (1989) "Reading like a feminist", *Differences* 1: 77–92.

Gage, N. (1989) *A Place for Us: Eleni's Children in America,* London: Black Swan.

Galinsky, E., Bond, J.T., and Friedman, D. (1993) *The Changing Workforce: Highlights of the National Study,* New York: Families and Work Institute.

Game, A. (1991) *Undoing the Social,* Toronto: University of Toronto Press.

Game, A. and Pringle, R. (1983) *Gender at Work,* Sydney: Allen and Unwin.

Garreau, J. (1991) *Edge City: Life on the New Frontier,* New York: Doubleday.

Gerstle, G. (1989) *Working Class Americanism: the Politics of Labour in a Textile City, 1914–1960,* Cambridge: Cambridge University Press.

Gilbert, M. (1993) Ties to people, bonds to place: the urban geography of low-income women's survival strategies. Ph.D. dissertation, Clark University.

—— (1994) "The politics of location: doing feminist research at 'home' ", *The Professional Geographer* 46: 90–6.

Giuliano, G. (1979) "Public transportation and the travel needs of women", *Traffic Quarterly* 33: 607–16.

Giuliano, G., Levine, D., and Teal, R. (1990) "Impact of HOV lanes on commuter travel behavior", *Transportation* 17: 159–77.

Glass, J. (1990) "The impact of occupational segregation on working conditions", *Social Forces* 68: 779–96.

Godfrey, K. and Kievra, B. (1990) "Valley awaiting bloom of growth: all depends on Mass Pike connector", *Worcester Telegram and Gazette* 18 December: 1.

Gordon, P., Kumar, A., and Richardson, H. (1989) "Gender differences in metropolitan travel behavior", *Regional Studies* 23: 499–510.

Granovetter, M. (1974) *Getting a Job: A Study of Contacts and Careers*, Cambridge, Mass.: Harvard University Press.

———— (1985) "Economic action and social structure: the problem of embeddedness", *American Journal of Sociology* 91: 481–510.

———— (1986) "Labor mobility, internal markets, and job matching: a comparison of the sociological and economic approaches", *Research in Social Stratification and Mobility* 5: 3–39.

———— (1988) 'The sociological and economic approaches to labor market analysis: a social structural view" in G. Farkas and P. England (eds.) *Industries, Firms, and Jobs: Sociological and Economic Approaches*, New York: Plenum.

Granovetter, M. and Tilly, C. (1988) "Inequality and the labor market", in N. Smelser (ed.) *Handbook of Sociology*, Beverly Hills: Sage.

Greater Worcester Multiple Listing Service, Inc. (1994) personal communication.

Gregory, D. (1990) "Chinatown, part three? Soja and the missing spaces of social theory", *Strategies: A Journal of Theory, Culture and Politics* 3: 40–104.

Grenier, G., Stepick, A., Draznin, D., LaBorwit, A., and Morris, S. (1992) "On machines and bureaucracy: controlling ethnic interaction in Miami's apparel and construction industries", in L. Lamphere (ed.) *Structuring Diversity: Ethnographic Perspectives on the New Immigration*, Chicago: University of Chicago Press.

Grieco, M. (1987) *Keeping it in the Family*, Cambridge: Polity Press.

Gross, E. (1968) "Plus ca change . . . the sexual segregation of occupations over time", *Social Problems* 16: 198–208.

Grosz, E. (1992) "Bodies–Cities", in B. Colomina (ed.) *Sexuality and Space*, New York: Princeton Architectural Press.

Hanson, S. (1992) "Geography and feminism: worlds in collision? ", *Annals of the American Association of Geographers* 82: 569–86.

Hanson, S. and Johnston, I. (1985) "Gender differences in work-trip length: explanations and implications", *Urban Geography* 6: 193–219.

Hanson, S. and Hanson, P. (1980) "Gender and urban activity patterns in Uppsala, Sweden", *Geographical Review* 70: 291–9.

———— (1981) "The impact of married women's employment on household travel patterns", *Transportation* 10: 165–83.

Hanson, S. and Pratt, G. (1988a) "Reconceptualizing the links between home and work in urban geography", *Economic Geography* 64: 299–321.

———— (1988b) "Spatial dimensions of the gender division of labor in a local labor market", *Urban Geography* 9: 180–202.

———— (1990) "Geographic perspectives on the occupational segregation of women", *National Geographic Research* 6: 376–99.

———— (1991) "Job search and the occupational segregation of women", *Annals of the American Association of Geographers* 81: 229–53.

———— (1992) "Dynamic dependencies: a geographic investigation of local labor markets", *Economic Geography* 68: 373–405.

Hanson, S., Pratt, G., Mattingly, D., and Gilbert, M. (1994) "Women, work, and metropolitan environments", in I. Altman and A. Churchman (eds.) *Women and the Environment*, New York: Plenum Press.

Haraway, D. (1991) "Situated knowledges: the science question in feminism and the privilege of partial perspective", in D. Haraway *Simians, Cyborgs, and Women: The Reinvention of Nature*, New York: Routledge.

Harrison, B. (1988) "Second thoughts on the Massachusetts miracle", *Technology Review* 91: 20 and 78.

Hartmann, H. (1979) "The unhappy marriage of Marxism and feminism: towards a more progressive union", *Capital and Class* 8: 1–33.

―――― (1987) "The family as the locus of gender, class and political struggle: the example of housework", in S. Harding (ed.) *Feminism and Methodology*, Bloomington and Indianapolis: Indiana University Press.

Henderson, Y. (1990) "Defense cutbacks and the New England economy", *New England Economic Review* July/August: 3–24.

Herwitz, E. (1987) "Area's labor pool reflects a mismatch of needs and resources", *Business Digest* February: 3ff.

Hill, J.A. (1929) *Women in Gainful Occupations, 1870 to 1920*, Census Monograph IX, U.S. Department of Commerce, Washington D.C.

Hirsch, B.T. and Macpherson, D.A. (1993) "Union membership: coverage files from the Current Population Surveys: note", *Industrial and Labor Relations Review* 46: 574–8.

Hirsch, M. and Fox Keller, E. (eds.) (1990) *Conflicts in Feminism*, New York: Routledge.

Holzer, H.J. (1991) "The spatial mismatch hypothesis: what has the evidence shown?", *Urban Studies* 28: 105–22.

hooks, b. (1992) "Representing whiteness in the black imagination", in L. Grossberg, C. Nelson and P. Treichler (eds.) *Cultural Studies* New York and London: Routledge.

―――― (1993) "Power to the pussy: we don't wannabe dicks in drag", in L. Frank and P. Smith (eds.) *Madonnarama: Essays on Sex and Popular Culture*, Pittsburgh: Cleis Press.

Howe, A. and O'Connor, K. (1982) "Travel to work and labor force participation of men and women in an Australian metropolitan area", *Professional Geographer* 34: 50–64.

Howland, M. (1988) *Plant Closings and Worker Displacement: The Regional Issues*, Kalamazoo, Michigan: Upjohn Institute.

Hughes, G.S. (1925) *Mothers in Industry: Wage Earning by Mothers in Philadelphia*, New York: New Republic.

Hughes, M. (1991) "Employment decentralization and accessibility: a strategy for stimulating regional mobility", *Journal of the American Planning Association* 57: 288–98.

Ihlanfeldt, K. and Young, M. (1994) "Intra metropolitan variation in wage rates: the case of Atlanta fast-food restaurant workers", *Review of Economics and Statistics*, in press.

Jacobs, J. (1989) *Revolving Doors: Sex Segregation and Women's Careers*, Stanford, California: Stanford University Press.

Jacobsen, J.P. and Levin, L.M. (1992) "The effects of intermittent labor force attachment on female earnings", paper given American Economic Association Conference, January 3–5, New Orleans, LA.

Jardine, A. (1985) *Gynesis: Configurations of Women and Modernity*, Ithaca and London: Cornell University Press.

Johnson, K. (1992) "Corporate elite a fading force over Hartford", *New York Times* 7 September: 1.

Johnston-Anumonwo, I. (1992) "The influence of household type on gender differences in work trip distance", *The Professional Geographer* 44: 161–9.

REFERENCES

—— (1994) Racial differences in the commuting behavior of women in Buffalo, 1980 to 1990. Mimeo.

Jonas, A. (1992) "Corporate takeover and the politics of community: the case of Norton Company in Worcester", *Economic Geography* 68: 348–72.

Jones-D'Agostino, S. (1987) "Valley whirl: the Blackstone Valley joins the Massachusetts Miracle", *Business Worcester* 6 September: 15.

Jones, J. and Rosenfeld, R. (1989) "Women's occupations and local labor markets: 1950 to 1980", *Social Forces* 67: 666–93.

Jordan, B., James, S., Kay, H., and Redley, M. (1992) *Trapped in Poverty? Labor Market Decisions in Low-Income Families*, London: Routledge.

Joshi, H. (1984) *Women's Participation in Paid Work: Further Analysis of the Women and Employment Survey*, London: Department of Employment.

Kain, J. (1968) "Housing segregation, negro employment, and metropolitan decentralization', *Quarterly Journal of Economics* 82: 175–97.

Kaletsy, A. (1988) "How miraculous is Massachusetts? ", *World Press Review* 35: 12.

Kay, H. (1990) "Research note: constructing the epistemological gap: gender divisions in social research", *Sociological Review* 38: 344–51.

Kishler Bennett, S. and Alexander, L.B. (1987) "The mythology of part-time work: empirical evidence from a study of working mothers", in C. Stimpson and L. Beneria (eds.) *Women, Households and the Economy*, New Brunswick and London: Rutgers University Press.

Kolesar, R.J. (1989) "The politics of development: Worcester, Massachusetts in the late nineteenth century", *Journal of Urban History* 16: 3–27.

Krefetz, S.P. (1992) "Urban economic revitalization as a catalyst for urban political change", in D.J. Etazar and Z. Marom (eds.) *Urban Revitalization*, New York: University Press of America.

Lakoff, G. and Johnson, M. (1980) *Metaphors we live by*, Chicago: University of Chicago Press.

Lamphere, L. (1987) *From Working Daughters to Working Mothers: Immigrant Women in a New England Community*, Ithaca, NY: Cornell University Press.

—— (ed.) (1992) *Structuring Diversity: Ethnographic Perspectives on the New Immigration*, Chicago: University of Chicago Press.

Lang, K. and Dickens, W. (1988) "Neoclassical and sociological perspectives on segmented labor markets", in G. Farkas and P. England (eds.) *Industries, Firms and Jobs: Sociological and Economic Approaches*, New York: Plenum.

Latour, B. (1987) *Science in Action*, Cambridge, Mass.: Harvard University Press.

Laue, L. (1988) "Massachusetts miracle", *Issues in Science and Technology* 5: 110.

Laurie, H. and Sullivan, O. (1991) "Combining qualitative and quantitative data in the longitudinal study of household allocations", *The Sociological Review* 39: 113–30.

Lawson, C. (1991) "Distance makes the heart skip for commuter moms: The perils of the modern Pauline", *New York Times* 7 November: C1, 2.

Ley, D. and Samuels, M. (1978) *Humanistic Geography*, Chicago: Maaroufa.

Lin, N. and Dumin, M. (1986) "Access to occupations through social ties", *Social Networks* 8: 365–85.

McDowell, L. (1991a) "Life without father and Ford", *Transactions, Institute of British Geographers* 16: 400–19.

—— (1991b) "The baby and the bathwater: diversity, deconstruction and feminist theory in geography", *Geoforum* 22: 123–33.

McDowell, L. and Massey, D. (1984) "A woman's place? ", in D. Massey and J. Allen (eds.) *Geography Matters!*, Cambridge: Cambridge University Press.

256

Mackenzie, S. (1989) "Women in the city", in R. Peet and N. Thrift (eds.) *New Models in Geography, vol. 2*, London: Unwin Hyman.

Mackenzie, S. and Rose, D. (1983) "Industrial change, the domestic economy and home life", in J. Anderson, S. Duncan and R. Hudson (eds.) *Redundant Spaces in Cities and Regions*, New York: Academic Press.

McLafferty, S. and Preston, V. (1991) "Gender, race, and commuting among service sector workers", *The Professional Geographer* 43: 1–14.

—— (1992) "Spatial mismatch and labor market segmentation for African-American and Latina women", *Economic Geography* 68: 406–31.

McRae, S. (1991) "Occupational change over childbirth: evidence from a national survey", *Sociology* 25: 589–605.

Madden, J.F. (1981) "Why women work closer to home", *Urban Studies* 18: 181–94.

Madden, J.F. and Chiu, L.C. (1990) "The wage effects of residential location and commuting constraints on employed married women", *Urban Studies* 27: 353–69.

Madden, J.F. and White, M. (1980) "Spatial implications of increases in the female labor force: a theoretical and empirical synthesis", *Land Economics* 56: 432–46.

Manwaring, T. (1984) "The extended internal labor market", *Cambridge Journal of Economics* 8: 161–87.

Maranz, M. (1987) "State of the unions", *Worcester Magazine* April: 9–12.

Massachusetts Institute for Social and Economic Research (MISER) (1994) Machine readable data tape 3A. Special run #1441, personal communication.

Massey, D. (1984) *Spatial Divisions of Labor: Social Structures and the Geography of Production*, New York: Methuen.

—— (1991) "Flexible sexism", *Environment and Planning D: Society and Space* 9: 31–57.

—— (1992) "A place called home?", *New Formations* 17: 3–15.

—— (1994) *Space, Place, and Gender*, Cambridge and Oxford: Polity Press.

Meagher, T.J. (1986) "Sweet good mothers and young women out in the world: the roles of Irish American women in later nineteenth and early twentieth century Worcester, Massachusetts", *U.S. Catholic Historian*, 5, 3–4: 325–44.

Medoff, J. (1994) personal communication. Calculations based on data from *Earning Differences between Women and Men*, Washington D.C.: Department of Labor, Women's Bureau.

Mier, R. and Giloth, R. (1985) "Hispanic employment opportunities: a case of internal labor markets and weak-tied social networks", *Social Science Quarterly* 66: 296–307.

Milkman, R. (1993) "Union responses to workforce feminization in the United States", in J. Jenson and R. Mahon (eds.) *The Challenge of Restructuring: North American Labor Movements Respond*, Philadelphia: Temple University Press.

Mincer, J. and Ofek, H. (1982) "Interrupted work careers: depreciation and restoration of human capital", *Journal of Human Resources* 17: 3–24.

Mincer, J. and Polachek, S. (1978) "An exchange theory of human capital and the earnings of women: women's earnings reexamined", *Journal of Human Resources*, 13: 118–34.

Mohanty, C.T. (1987) "Feminist encounters: locating the politics of experience", *Copyright* 1: 30–44.

Montgomery, J.D. (1991) "Social networks and labor market outcomes: towards an economic analysis", *American Economic Review* 81: 1408–18.

REFERENCES

Moore, G. (1990) "Structural determinants of men's and women's personal networks", *American Sociological Review* 55: 726–35.

Moore, H. (1988) *Feminism and Anthropology*, Minneapolis: University of Minnesota Press.

Morris, L. (1990) *The Workings of the Household: A U.S.–U.K. Comparison*, Cambridge: Polity Press.

—— (1994) "Informal aspects of social divisions", *International Journal of Urban and Regional Research* 18: 112–26.

Morrison, P.S. (1990) "Segmentation theory applied to local, regional and spatial labour markets", *Progress in Human Geography* 14: 488–528.

Moscovitch, E. (1990) "The downturn in the New England economy: what lies behind it?", *New England Economic Review*, June/July: 53–65.

Moseley, M.J. and Darby, J. (1978) "The determinants of female activity rates in rural areas: an analysis of Norfolk parishes", *Regional Studies* 12: 297–309.

Mulligan, W.H. Jr. (1980) "Divorce in Worcester County, Massachusetts, 1863–1880", *Journal of Family Issues* 1: 357–75.

Myers, S. (1970) "Personal transportation for the poor", *Traffic Quarterly* 24, 191–206.

Nakano-Glenn, E. (1985) "Racial ethnic women's labor: the intersection of race, gender, and class oppression", *Review of Radical Political Economics* 17: 86–108.

Nelson, K. (1986) "Female labor supply characteristics and the suburbanization of low-wage office work", in M. Storper and A. Scott (eds.) *Production, Work, and Territory*, Boston: Allen and Unwin.

Newton, K. (1985) "Chamber officer sparks belief in Valley plans", *Worcester Telegram and Gazette*, 3 December: 21B.

Nutt, C. (1919) *History of Worcester and its People*, New York: Lewis Historical Publishing Company.

O'Connell, M. (1990) "Maternity leave arrangements, 1961–1985", in *Work and Family: Patterns of American Women*, Washington D.C.: U.S. Department of Commerce.

Okely, J. and Callaway, H. (eds.) (1992) *Anthropology and Autobiography*, London: Routledge.

Oppenheimer, V.K. (1973) "Demographic influence on female employment and the status of women", *American Journal of Sociology* 78: 946–61.

Parr, J. (1990) *The Gender of Breadwinners: Women, Men and Change in Two Industrial Towns, 1880–1950*, Toronto: University of Toronto Press.

Pas, E. (1984) "The effect of selected sociodemographic characteristics on daily travel behavior", *Environment and Planning A* 16: 571–81.

Peck, J. (1989) "Reconceptualizing the local labor market: space, segmentation, and the state", *Progress in Human Geography* 13: 42–61.

Perry, S. (1988) "Downward occupational mobility and part-time women workers", *Applied Economics* 20: 485–95.

—— (1990) "Part-time work and returning to work after the birth of the first child", *Applied Economics* 22: 1137–48.

Personal Narratives Group (1989) *Interpreting Women's Lives: Feminist Theory and Personal Narratives*, Bloomington: Indiana University Press.

Peterson, B.R. (1989) "Firm size, occupational segregation, and the effect of family status on woman's wages", *Social Forces* 68: 397–414.

Phillips, A. (1987) *Divided Loyalties: Dilemmas of Sex and Class*, London: Verso.

Pinch, S. (1987) "Labour-market theory, gentrification and policy", *Environment and Planning A* 19: 1477–94.

REFERENCES

Pinch, S. and Storey, A. (1992) "Who does what where?: a household survey of the division of labour in Southampton", *Area* 24: 5–12.

Pleck, J. (1985) *Working Wives, Working Husbands*, Beverly Hills, CA: Sage Publications.

Polachek, S. (1981) "Occupational self-selection", *Review of Economics and Statistics* 58: 60–9.

Pollock, G. (1988) *Vision and Difference: Femininity, Feminism and Histories of Art*, New York: Routledge.

Pope, C. (1994) "Ownership shifting from city", *Worcester Telegram and Gazette* 13 May: A7.

Pratt, E.E. (1911) *Industrial Causes of Congestion of Population in New York City*, New York: Columbia University Press.

Pratt, G. (1993) "Geographic metaphors in feminist theory", paper presented at "Making Worlds: Metaphor and Materiality in the Production of Feminist Texts Conference", October 14–16, Tucson, Arizona.

Pratt, G. and Hanson, S. (1988) "Gender, class, and space", *Environment and Planning D: Society and Space* 6: 15–35.

—— (1990) "On the links between home and work: family strategies in a buoyant labor market", *International Journal of Urban and Regional Research* 14: 55–74.

—— (1991) "Time, space, and the occupational segregation of women: a critique of human capital theory", *Geoforum* 22: 149–57.

—— (1993) "Women and work across the life course: moving beyond essentialism", in C. Katz and J. Monk (eds.) *Full Circles: Geographies of Women Over the Life Course*, New York: Routledge.

—— (1994) "Geography and the construction of difference", *Gender, Place, and Culture* 1: 5–29.

Pugh, A. (1990) "My statistics and feminism – a true story", in E. Stanley (ed.) *Feminine Praxis*, London: Routledge, pp. 103–12.

Purcell, W. (1989) personal communication, Executive Vice-President, Worcester Business Development Corporation, May.

Redclift, N. and Sinclair, M.T. (eds.) (1991) *Working Women: International Perspectives on Labour and Gender Ideology*, London: Routledge.

Rees, A. and Schultz, G.P. (1970) *Workers and Wages in an Urban Labor Market*, Chicago: University of Chicago Press.

Research Department of the Women's Educational and Industrial Union of Boston (1913) "A Trade School for Girls: A Preliminary Investigation in a Typical Manufacturing City, Worcester, Massachusetts", pamphlet, Washington D.C.: Government Printing Office.

Reskin, B. and Hartmann, H. (1986) *Women's Work, Men's Work: Sex Segregation on the Job*, Washington D.C.: National Academy Press.

Reskin, B. and Roos, P. (1990) *Job Queues and Gender Queues: Explaining Women's Inroads into Male Occupations*, Philadelphia: Temple University Press.

Reynolds, D. (1989) "Shifting gears: the changing work experience of Blackstone Valley workers, 1920–1989", pamphlet, Massachusetts Foundation for the Humanities and Public Policy.

—— (1991) "Deindustrialization in the Blackstone Valley: 1920–1989", in D. Reynolds and M. Meyers (eds.) *Working in the Blackstone River Valley: Exploring the Heritage of Industrialization*, Woonsocket: Rhode Island Labor History Association.

Rich, A. (1986) *Blood, Bread, and Poetry: Selected Prose, 1979–1985*, New York: Norton.

259

REFERENCES

Rice, F.B. (1898) *The Dictionary of Worcester, Massachusetts and Surrounding Vicinity*, Worcester: F.S. Blanchard.

Roberts, H. (1984) "The dissemination of research findings", in C. Bell and H. Roberts (eds.) *Social Researching: Politics, Problems and Practice*, London, Boston: Routledge and Kegan Paul, pp. 191–212.

Robins, K. (1991) "Tradition and translation: national culture in its global context", in J. Corner and S. Honey (eds.) *Enterprise and Heritage*, London: Routledge.

Rose, D. (1993) "Local childcare strategies in Montreal, Quebec", in C. Katz and J. Monk (eds.) *Full Circles: Geographies of Women over the Life Course*, London: Routledge.

Rose, G. (1993) *Feminism and Geography: the Limits of Geographical Knowledge*, Cambridge: Polity Press.

Rosenberg, R. (1991) "Biotech West", *Boston Globe* 23 July: 33.

Rosenbloom, S. and Burns, E. (1993) "Gender differences in commuter travel in Tucson: implications for travel demand management programs", *Transportation Research Record* #1404: 82–90.

Rosenfeld, R.A. (1979) "Women's occupational careers: individual and structural explanations", *Sociology of Work and Occupations* 6: 283–311.

Rosenzweig, R. (1983) *Eight Hours for What We Will: Workers and Leisure in an Industrial City, 1870–1920*, New York: Cambridge University Press.

Ross, R.J.S. and Trachte, K.C. (1990) *Global Capitalism: The New Leviathan*, New York: SUNY Press.

Rubery, J. (1978) "Structured labour markets, worker organisation and low pay", *Cambridge Journal of Economics* 2: 17–36.

Rytina, N.F. and Bianchi, S. (1984) "Occupational reclassification and changes in distribution by gender", *Monthly Labor Review* 107: 11–17.

Sacks, K.B. (1989) "Toward a unified theory of class, race, and gender", *American Ethnologist* 16: 534–50.

Saegert, S. (1980) "Masculine cities and feminine suburbs: polarized ideas, contradictory realities", *Signs: Journal of Women in Culture and Society* 5: S-96-S-111.

Salopek, P. (1992) personal communication, 19 November.

Saxenian, A. (1994) *Regional Advantage: Culture and Competition in Silicon Valley and Route 128*, Cambridge: Harvard University Press.

Schantz, S. (1991) "Please be mine", *Inside Worcester* February: 33–5.

Schor, N. (1989) "This essentialism which is not one: coming to grips with Irigaray', *Differences* 1: 38–58.

Schroedel, J.R. (1985) *Alone in a Crowd: Women in the Trades Tell Their Stories*, Philadelphia: Temple University Press.

Scott, A.J. (1988) *Metropolis: From the Division of Labor to Urban Form*, Berkeley: University of California Press.

—— (1992a) "Low-wage workers in a high-technology manufacturing complex: the Southern California electronics assembly industry", *Urban Studies* 29: 1231–46.

—— (1992b) "The spatial organization of a local labor market: employment and residential patterns in a cohort of engineering and scientific workers", *Growth and Change* 23: 94–115.

Sege, I. (1989) "Tough future in N.E. calls for ingenuity", *Boston Globe* 12 September: 1.

Sheppard, E. and Barnes, T. (1990) *The Capitalist Space Economy*, London: Unwin Hyman.

Siltanen, J. (1986) "Domestic responsibilities and the structuring of employment", in R. Crompton and M. Mann (eds.) *Gender and Stratification*, Cambridge: Polity Press.

Sinacola, C. (1991) "Valley positions itself for better times", *Worcester Telegram and Gazette* 15 December: 1.

Sinclair, M.T. (1991) "Women, work, and skill: economic theories and feminist perspectives", in N. Redclift and M.T. Sinclair (eds.) *Working Women: International Perspectives on Labour and Gender Ideology*, London: Routledge.

Singell, L. and Lillydahl, J. (1986) "An empirical analysis of the commute to work pattern of males and females in two-earner households", *Urban Studies* 23: 119–29.

Sit, M. (1991) "Fringe benefits: some companies are helping employees with housing", *Boston Globe* 20 July: 37.

Smith, D.E. (1990) *Text, Facts and Femininity: Exploring the Relations of Ruling*, London: Routledge.

Smith, J. (1987) "Transforming households: working-class women and economic crisis", *Social Problems* 24: 416–36.

Smith, P. (1988) *Discerning the Subject: Theory and History of Literature, vol. 25*, Minneapolis: University of Minnesota Press.

Soja, E. (1989) *Postmodern Geographies: The Reassertion of Space in Critical Social Theory*, London: Verso.

Sommers, C.H. (1994) *Who Stole Feminism?*, New York: Simon and Schuster.

Southwick, A.B. (1988) 'Don't look for the union label", *Worcester Monthly* December: 22–9.

Spain, D. (1992) *Gendered Spaces*, Chapel Hill: University of North Carolina Press.

Spalter-Roth, R. and Hartmann, H. (1991) "Improving women's status in the workforce: the family issues of the future", testimony given U.S. Senate, Women and the Workplace: Looking Toward the Future, Hearings of the Subcommittee on Employment and Productivity Committee on Labor and Human Resources, July 18.

Spear, M. (1977) "Worcester's Three-deckers", pamphlet, Worcester, MA: Worcester Bicentennial Commission.

Spivak, G.C. (1983) "Displacement and the discourse of woman", in M. Krupnick (ed.) *Displacement: Derrida and After*, Bloomington: Indiana University Press

—— (1990) "Strategy, identity, writing, in S. Harasmyn (ed.) *The Postcolonial Critic: Interviews, Strategies, and Dialogues*, New York: Routledge, Chapman and Hall.

—— (1993) *Outside in the Teaching Machine*, New York and London: Routledge.

Stacey, J. (1990) *Brave New Families: Stories of Domestic Upheaval in the Late Twentieth Century America*, United States: Basic Books.

Stanton, E.C., Anthony, S.B., and Gage, M.J. (1881) *History of Woman Suffrage, Vol. 1: 1848–1861*, reprint ed. 1969, New York: Arno and the New York Times.

Statistics Canada (1993) *Labour Force Activity of Women by Presence of Children*, Ottawa: Ministry of Industry, Science and Technology.

Stein, C. (1993) "Grim outlook beyond Route 128", *Boston Globe* 27 July: 33.

Steinbruner, M. and Medoff, J. (1994) *Jobs and the Gender Gap: The Impact of Structural Change on Worker Pay, 1984–1993*, Washington, D.C.: Center for National Policy.

Stevens, D. (1978) "A reexamination of what is known about job seeking behavior in the U.S." in *Labor Market Intermediaries Special Report #22*, Washington, D.C.: The National Commission for Manpower Policy.

Stevens, G. and Cho, J.H. (1985) "Socioeconomic indices and the new 1980 census occupational classification scheme", *Social Science Research* 15: 142–68.

Stolzenberg, R. and Waite, L. (1984) "Local labor markets, children, and labor force participation of wives", *Demography* 21: 157–70.

Storper, M. and Walker, R. (1983) "The theory of labor and the theory of location", *International Journal of Urban and Regional Research*, 7: 1–43.

Stull, D., Broadway, M.J. and Erickson, K. (1992) "The price of a good steak: beef packing and its consequences for Garden City, Kansas", in L. Lamphere (ed.) *Structuring Diversity: Ethnographic Perspectives on the New Immigration*, Chicago: University of Chicago Press.

Thrift, N. and Williams, P. (1987) *Class and Space: the Making of Urban Society*, London: Routledge and Kegan Paul.

Tivers, J. (1985) *Women Attached: The Daily Lives of Women with Young Children*, London: Croom Helm.

Tomaskovic-Devey, D. (1993) *Gender and Racial Inequality at Work: the Sources and Consequences of Job Segregation*, Ithaca, New York: ILR Press.

Tong, R. (1989) *Feminist Thought: A Comprehensive Introduction*, Boulder: Westview Press.

Towne, J. (1979) "Women face struggle looking for jobs", *Worcester Telegram and Gazette* 5 September: 1D.

Trinh, M. (1990) "Cotton and iron", in R. Ferguson, M. Gever, M.T. Trinh, and C. West (eds.) *Out There: Marginalization and Contemporary Culture*, New York and Cambridge, MA: The New Museum of Contemporary Art and MIT Press.

Troy, L. and Sheflin, N. (1985) *Union Sourcebook: Membership, Structure, Finance Directory*, West Orange, N.J.: Industrial Relations Data Information Services.

U.S. Census Office (1890) *Population of the United States at the Eleventh Census*, Washington, D.C.: U.S. Department of the Interior.

—————— (1900) *Population of the United States at the Twelfth Census*, Washington, D.C.: U.S. Department of the Interior.

—————— (1910) *Population of the United States at the Thirteenth Census*, Washington, D.C.: U.S. Department of the Interior.

U.S. Bureau of the Census (1920) *Fourteenth Census of the United States*, Washington, D.C.: U.S. Department of Commerce.

—————— (1930) *Fifteenth Census of the United States*, Washington, D.C.: U.S. Department of Commerce.

—————— (1940) *Sixteenth Census of the United States*, Washington, D.C.: U.S. Department of Commerce.

—————— (1950) *Census of Population: 1950*, Washington, D.C.: U.S. Department of Commerce.

—————— (1960) *Census of Population: 1960*, Washington, D.C.: U.S. Department of Commerce.

—————— (1962) *County and City Data Book*, Washington, D.C.: U.S. Department of Commerce.

—————— (1970) *1970 Census of Population*, Washington, D.C.: U.S. Department of Commerce.

—————— (1980) *1980 Census of Population and Housing*, Washington, D.C.: U.S. Department of Commerce.

—————— (1983) *Census of Population and Housing, 1980: Public-Use Microdata Samples, Technical Documentation*, Washington, D.C.: U.S. Department of Commerce.

REFERENCES

———— (1990a) *1990 Census of Population and Housing*, Washington, D.C.: U.S. Department of Commerce.

———— (1990b) *Census of Population and Housing, 1990: Public-Use Microdata Samples, Technical Documentation*, Washington, D.C.: U.S. Department of Commerce.

———— (1993) *Statistical Abstract of the United States*, Washington, D.C.: U.S. Department of Commerce.

Valentine, G. (1989) "The geography of women's fear", *Area* 21: 385–90.

Vancouver Sun 9 May 1994: A1.

Vipond, J. (1984) "The intra-urban unemployment gradient: the influence of location on unemployment", *Urban Studies* 21: 377–88.

Virilio, P. (1986) "The Overexposed City", *Zone* 1/2: 14–39.

Wahlstrom, E.M. (1947) "A History of the Swedish people of Worcester, Massachusetts", unpublished M.A. thesis, Clark University.

Wajcman, J. (1991) *Feminism Confronts Technology*, University Park, PA: Penn State University Press.

Walby, S. (1986) *Patriarchy at Work*, Cambridge: Polity Press.

———— (1989) "Theorising patriarchy", *Sociology* 23: 213–34.

Walby, S. and Bagguley, P. (1989) "Gender restructuring: five labour markets compared", *Environment and Planning D: Society and Space* 7: 277–92.

Walker, R.A. (1985) "Class, division of labor and employment in space", in D. Gregory and J. Urry (eds.) *Social Relations and Spatial Structures*, New York: St. Martin's Press.

Ward, C. and Dale, A. (1992) "Geographical variation in female labor force participation: an application of multilevel modelling", *Regional Studies* 26: 243–55.

Warren, S. (1987) "Housing and ideology: the menace of the three-decker", unpublished M.A. thesis, Clark University.

Warshaw, M. (1991) "Little Saigon, Central Mass.", *Inside Worcester* May: 12–19.

Wekerle, G. and Rutherford, B. (1989) "The mobility of capital and the immobility of female labor: responses to economic restructuring", in J. Wolch and M. Dear (eds.) *The Power of Geography*, Boston: Unwin Hyman.

Wheelock, J. (1990) *Husbands at Home: the Domestic Economy in a Post-Industrial Society*, London: Routledge.

Whipp, R. (1985) "Labour market and communities: an historical view", *Sociological Review* 19: 768–91.

White, J. (1983) *Women and Part-time Work*, Ottawa: Ministry of Supply and Services.

Wial, H. (1991) "Getting a good job: mobility in a segmented labor market", *Industrial Relations* 30: 396–416.

Wilson, E. (1991) *The Sphinx in the City: Urban Life, the Control of Disorder, and Women*, Virago: London and Berkeley: University of California Press.

Wolff, J. (1985) "The invisible flaneuse: women and the literature of modernity", *Theory, Culture and Society* 2: 37–48.

———— (1990) *Feminine Sentences: Essays on Women and Culture*, Berkeley: University of California Press.

———— (1993) "On the road again: metaphors of travel in cultural criticism", *Cultural Studies* 7: 224–39.

Woolgar, S. (1988) *Science, the very idea*, London: Tavistock.

Worcester Telegram (1965) "City-owned land made available for new Sprague Electric Co. plant", 19 June.

Worcester Telegram (1986) "Why was the toll road built so far south? ", 17 March: 12.

REFERENCES

Young, I.M. (1989) "Throwing like a girl", in J. Allen and I.M. Young (eds.) *The Thinking Muse*, Bloomington: Indiana University Press.

Zeuch, W. (1916) "An investigation of the metal trades strike of Worcester, 1915", unpublished M.A. thesis, Clark University.

INDEX

Note: Numbers in italic refer to figures or plates

265